T0325120

Education and Training for the Oil and Gas Industry: The Evolution of Four Energy Nations

Education and Training for the Oil and Gas Industry: The Evolution of Four Energy Nations

Mexico, Nigeria, Brazil and Iraq

VOLUME 3

Jim Playfoot
Phil Andrews
Simon Augustus

ELSEVIER

AMSTERDAM • BOSTON • HEIDELBERG • LONDON • NEW YORK • OXFORD
PARIS • SAN DIEGO • SAN FRANCISCO • SINGAPORE • SYDNEY • TOKYO

Elsevier
Radarweg 29, PO Box 211, 1000 AE Amsterdam, Netherlands
The Boulevard, Langford Lane, Kidlington, Oxford OX5 1GB, UK
225 Wyman Street, Waltham, MA 02451, USA

Notices
Knowledge and best practice in this field are constantly changing. As new research
and experience broaden our understanding, changes in research methods, professional
practices, or medical treatment may become necessary.

Practitioners and researchers must always rely on their own experience and knowledge
in evaluating and using any information, methods, compounds, or experiments
described herein. In using such information or methods they should be mindful of
their own safety and the safety of others, including parties for whom they have a
professional responsibility.

To the fullest extent of the law, neither the Publisher nor the authors, contributors, or
editors, assume any liability for any injury and/or damage to persons or property as a
matter of products liability, negligence or otherwise, or from any use or operation of
any methods, products, instructions, or ideas contained in the material herein.

ISBN: 978-0-12-800974-1

British Library Cataloguing in Publication Data
A catalogue record for this book is available from the British Library

Library of Congress Cataloging-in-Publication Data
A catalog record for this book is available from the Library of Congress

For Information on all Elsevier publications
visit our website at http://store.elsevier.com/

Working together
to grow libraries in
developing countries

www.elsevier.com • www.bookaid.org

Dedication

This book is dedicated to all those who are committed to building sustainable, equitable and accessible education and training systems around the world.

Contents

About the Authors

JIM PLAYFOOT

Jim is a consultant, researcher and writer working in the field of education and skills development. He is Founder and Managing Director of the London-based education consultancy White Loop.

Jim's work over the last 10 years has focussed on understanding the dynamic between education and employment, exploring the challenges of how we prepare young people for the twenty-first century and developing new thinking around how education can have real impact on wellbeing and quality of life. Jim's work is built around a deep understanding of how people learn, allied to an ability to engage, analyse and interpret evidence and opinion and produce outputs that are compelling and accessible. He has worked with partners and collaborators in more than 20 countries around the world and is increasingly focussed on exploring the potential for education as a means of promoting development objectives in Africa.

In 2011, he was approached by Getenergy Events to develop their research and intelligence function. In 2014, Jim was part of the team that established Getenergy Intelligence and assumed the role of Managing Director, a position that will see him lead the authoring of all four of the Getenergy Guides volumes.

PHIL ANDREWS

Most of Phil's career has been as an entrepreneur connected to education and training in the upstream oil and gas business. Having graduated from Nottingham University with a degree in politics he was made an Honorary Life Member of the University Union. In 2003, after some travel and a short period working in conferences and publishing, Phil became half of the partnership that founded Getenergy Events in Aberdeen. He has since been awarded the Livewire Young Entrepreneur of the Year Award in London and was runner-up in the UK-wide competition.

Under Phil's leadership, Getenergy's brand has achieved widespread recognition in the oil and gas and education sectors and more than 40 countries are regularly involved in its meetings, networks and publications. Phil is often asked to speak at international events and is an influential and well-respected international figure in the oil and gas industry who has visited more than 45 countries in support of Getenergy's activities including Iraq, Libya and Equatorial Guinea. His opinion is increasingly sought by Ministers and their governments in matters connected to developing sustainable economic value through training and education funded by oil and gas activities.

SIMON AUGUSTUS

Simon brings his deep knowledge and understanding of the global energy and industrial gas industries to his role as Lead Analyst at Getenergy Intelligence. He has written extensively in these areas and contributed as a regular feature writer to several trade magazines within the oil, natural gas and renewable energy sectors.

Prior to joining Getenergy Intelligence in 2014, Simon worked with a range of international clients, including the governments of Turkey and Egypt. His work today is particularly concerned with exploring the dynamic between the oil and gas industry and the wider geopolitical context alongside a focus on economic development and sustainable growth. He has a special interest in the economy and politics of Russia and the Middle East.

Preface

Welcome to the third book in the Getenergy Guides series. In this volume, we have chosen to look in detail at the evolution of four energy nations: Brazil, Mexico, Nigeria and Iraq. It is our belief that by analysing the efforts of others, we can collectively develop a deeper understanding of how to approach the complex challenges of building education and training systems that work for the people and that utilise revenues from—and investments in—hydrocarbon exploration and production to achieve this goal. Our aim here is simple: to learn from both the successes and the mistakes of others so that, together, we can chart a path towards a more sustainable, equal and productive model of how a country should respond when they find that they have a winning ticket in the global geological lottery.

The countries we explore in this volume have been chosen above others because each offers us a different—but comparable—story. We have tried to look beyond commonly held perspectives and to reach a more nuanced understanding of how these countries have engaged with the challenges and opportunities presented by the discovery of commercial quantities of oil and gas and what this might mean for education, skills and workforce development. As always, there are many other stories we could have chosen to tell (and we hope to do that in future editions).

By way of introducing this, our third volume, we would like to invite you to consider the plight of two countries not covered in this book. For reasons that will become clear later, let's call them Country A and Country B.

The story of Country A starts 20 years ago when commercial quantities of gas were discovered offshore. The discovery was significant enough to have the potential to change the economic underpinning of the country, to transform export revenues and improve the social, educational and employment prospects of many of the country's 20 million citizens (most of whom were poor and employed in the informal sector). At the point where the gas was discovered, the government of the day recognised the enormous opportunity that this presented. Here was the chance to generate significant revenues that could provide the impetus for major investments in infrastructure, education, health and the wider economy. In a country struggling with massive human development challenges, this was a game changer.

From the moment the first exploratory wells were drilled, it was clear that the government had one principal objective: get the gas out of the ground as quickly as possible. The political pressures were vast: when the value of gas reserves was set against the social and economic problems under which the country was labouring, the primary objective was to get the gas—and the subsequent revenues—flowing. Political leaders, wary of the 'resource curse', recognised the need to do more than simply sell off the concessions to the highest international bidder. They needed to implement a policy of local content that would ensure local participation across the sector. However, they also needed to be pragmatic; their basic education system was one of the least effective in the world and there was simply nothing in place to educate and train people to assume positions in what is a highly technical industry (and one where safety and productivity are paramount). Yes, local participation had to be the aim but not if it meant delaying the flow of revenue.

In response to these challenges (and in absence of a national oil company or a local commercial oil and gas sector) the government decided to invite international companies to bid for available concessions. At the same time, they would implement a robust local content policy that would require those bidding companies to work towards local participation over time. As the revenues began to accrue from production, the government could then start to invest in primary and secondary education, build new tertiary colleges and create the basis for the development of a new local workforce supported by the actions and initiatives of the international companies who set up locally. However, things did not go according to plan. The local content policy, while worthy in intention, failed to become a legal requirement for the first decade of production activities. Political mismanagement and a fear of scaring off international operators ensured that the idea of local participation remained unrealised. In the meantime, the industry expanded at a rapid rate. International operating and service companies poured into the country, bringing their own expertise, their own processes and their own approaches to workforce development. Without a coherent workforce development strategy for the country, these companies recruited and trained according to their own short-term needs. This typically meant bringing in expatriate workers from overseas or, at best, sending a small number of local recruits to overseas training centres to develop their skills.

Ten years in and the industry was safe and productive and the government was reaping the benefits of billions of dollars in export revenues. However, there were mounting problems. As the workforce was largely made up of imported workers, the local communities around exploration and production sites were increasingly unhappy at what they perceived as a 'foreign industry' capitalising on their natural resources without offering them anything in return (apart from the occasional environmental disaster). The international operators, for their part, saw no value and no opportunity to invest locally, as the skills base was simply not there. The challenge of educating locals seemed so vast as to be insurmountable. And anyway, they were all now well-established and their operations were running safely and productively. Why change?

Fast forward a further 10 years and the situation remains unchanged. Despite some of the wealth filtering down to communities in the form of big infrastructure projects and the construction of some new schools and hospitals, the real economic benefits of the industry have been concentrated amongst a small number of government figures who have been well-placed to channel revenues into the hands of friends, relations and associates. The other winners, of course, have been the international companies operating with impunity for two decades. But the real cost of the approach is more than just the unfair distribution of energy wealth: in spite of the opportunity that presented itself, there is no new skills base in Country A, there has been no significant investment in local education and training capacity, there are no national industrial leaders working in the sector, there is no local control and no legacy of skills or education to speak of. The country has become an economic slave to fluctuations in global energy prices. And worse still, the real tragedy of this story is yet to play out: Country A will run out of gas in 10 years' time. In this post-natural gas period what challenges will await its citizens? What will it have gained from three decades of being a 'hydrocarbon nation'? How will history judge the custodians of this natural resource gift?

So to Country B: Country B is similar in many ways to Country A. Its journey began with the discovery of large deposits of onshore oil around the same time as Country A discovered gas. Like Country A, Country B was economically under-developed. The political and economic pressures were very similar. Faced with the opportunity to generate significant revenues to bolster an ailing agrarian economy, the 'rush to revenue' was tempting. However, Country B, under the guidance of astute political leadership, had the vision and the courage to think differently. The vision was simple: to create a new economy and a new skills base from both the employment opportunities that the oil sector offered and from the revenues and investment it stimulated. Courage was needed to plan for the long term and resist the temptation to subjugate the vision on the altar of quick money.

The first step was to create a strategy that would oversee two things: the way in which the industry would operate and the connection to skills, education and workforce development. On the first point, it was recognised that Country B could not simply go it alone. For the industry to thrive in a country that had no history of oil production, there needed to be collaboration with international companies. However, that relationship would have to be one of equals. The government of Country B quickly established a national oil company able to explore and operate across the country alongside a regulatory body that would oversee the participation of commercial entities (including, but not restricted to, international operators). In those early days, the message to international companies was welcoming and positive but also forthright and cautionary: we want you to come and work with us but only if you are committed to genuine partnership and to building local capacity. Through the enacting of a coherent local content policy, the establishment of a regulator who had the teeth to enforce this policy and the investment in a national oil company as a partner in production sharing agreements, Country B created the foundations for an equitable and sustainable industry. At the same time, the government invested heavily in understanding and then building the local supply chain. Aware as it was of the limited number of direct jobs in the oil sector, there needed to be much greater focus on indirect employment and on ensuring that local companies were equipped and competitive enough to win that business.

So to the second part of the strategy: the connection to skills, education and workforce development. There were a number of aspects to this. First, the government realised at an early stage that a local content policy would mean nothing without the skills development activities to back it up. How can you expect national or international companies to employ local people if they don't possess the right knowledge, skills and behaviours for the job? If the oil sector in Country B was to become a driver for local employment – both directly and indirectly – then local people would need to be educated and trained in order to fulfil those roles. The government set out to do two things in response to this. First, it launched a skills audit in order to identify whether there were already candidates in the informal economy or in related sectors that could be retrained for the oil industry. Second, it created a map of the skills demands covering every aspect of the oil sector value chain. This meant understanding exactly what the skills demands would be during the construction phase, who would be needed once the plants were operational and what the demands would be in related sectors (including midstream, downstream, logistics, transport, maintenance, shipping, hospitality and so on). By mapping the supply and setting that against anticipated demand, a clear picture of the gap emerged. This formed the basis for the next step.

Knowing the skills you need to develop is one thing; actually doing it is another entirely. The approach in Country B would have to address this, and fast, but would also need to be commensurate with the initial vision. What would be the point of creating short-term sticking plasters for skills gaps when the opportunity was there for long-term healing of the whole education and training system? Country B set to work, driven by the principle that education and skills development should be delivered locally. Following a further audit of the education and training landscape—identifying the local colleges, universities and training providers that could possibly be part of the solution—the government created a plan based on building local capacity and meeting the needs of industry. This plan was developed in partnership with international oil companies, local and international education and training providers and the relevant government departments. At its heart was a clear mechanism to build the participation of industry into the design and delivery of education and training provision. The government also recognised the importance of meeting international standards for education and training and the role that the global education and training community could play in building this in-country. Furthermore, the government developed a wider plan that encompassed investments into basic education, a mapping of skills demands in other sectors and a plan for the transferability of knowledge, skills and competencies into new and nascent areas of the economy. This was not simply about building a workforce for the oil sector –

it was about creating a system of education and training that could sustain the economic ambitions of the country for generations to come.

The final piece of the puzzle was funding—the government committed to channelling a significant proportion of oil revenues into education and training facilities, infrastructure and programmes and into building the human capabilities of the teachers, trainers and managers who would drive forward this new system. They also imposed a levy on all commercial activities that would provide further economic ballast to support system development. Alongside this, they implemented a national framework for qualifications that was clearly defined and mapped against international standards creating the basis of a competitive private education and training sector that would ultimately thrive without the financial support of government or industry.

Today, 20 years on, what has Country B achieved? Well, it has an oil sector that is dominated at every level by nationals who have been locally trained and educated. It has a thriving education and training system that is responsive to industry needs, that is internationally benchmarked and that is well-respected across the world. It has a population of people who are deeply engaged with the economic success that oil has brought and that now have the skills and mobility to compete for employment opportunities and for contracts both at home and overseas. There is still some way to go, but the country has been transformed. Oil has been converted into skills.

Oh, and one more thing: Country B will also run out of oil in 10 years' time. However, the country is not having to deal with the difficult questions facing Country A. Its hydrocarbon legacy will be a completely reborn education and training system that is already developing citizens who are able to build a sustainable economic future today and will be well-placed to do so after the oil runs out. This investment in education and skills has helped the economy diversify—into green energy, manufacturing, engineering, advanced agriculture and so on. Country B is not reliant on diminishing and irreplaceable natural resources—it is facing a future as an economic powerhouse driven by the knowledge and creativity of its citizens. And it is education that has made this possible.

So what do we learn from a comparison of Country A and Country B? First, we can see how the gift of natural resources can, in fact, be a curse. We can also see how transformative such a gift can be if handled in the right way. Can we glean any truths from these two stories? We suggest the following:

- That to succeed takes vision and the political will to see that vision through
- That short-term thinking is the enemy of meaningful education reform
- That the notion of national control and national participation should be tempered by a realisation that international partnerships can, if managed in the right way, help to create local capacity and build the local workforce
- That the debate about skills and workforce development needs to go beyond a narrow focus on the requirements of a single sector; the whole system should be reformed for the good of all

- That understanding what a country needs in terms of skills is vital and that this goes way beyond the direct jobs that will accrue from hydrocarbon production and exploration activities
- That the effective implementation of policies aimed at promoting local participation is as critical as the policy itself
- That creating a legacy that outlives the oil and gas that is in the ground will only be achievable if education and training capacity is developed at a local level

In the four stories we present in this volume, you will recognise many of the themes identified in our (somewhat brief) exploration of Country A and Country B. In each of the four stories, we imagine the intentions of the main protagonists to be largely positive. However, as you will see, it is never easy. Whilst we commend the efforts of successive governments, international operators, educators and others, we also need to cast a critical eye on their failures. We do this only in an attempt to learn and to then share this learning with our readers.

So to Country A and Country B: The observant (or perhaps the cynical) amongst you may have already guessed that Country A and Country B do not exist, at least not literally. Rather we have created these imaginary states to represent opposite sides of the same coin, taking the best and worst of what we know and what we have been told. Simply put, Country A is the nightmare and, for us, Country B is the dream. Let's hope that by working together, building a shared vision and learning from the triumphs and failures of others, we can make the dream a reality in every oil- and gas-producing country in the world.

Jim Playfoot, Phil Andrews and Simon Augustus
Getenergy Intelligence
April 2015

Acknowledgements

We would like to thank the Getenergy team in London for their support in producing this book: Helen Jones, Virginia Baker, Jack Pegram, Kelly Hutchinson, Richard Harmon, Annameka Porter-Sinclair, Nick Cressey, Concepción Perez, Tom Fraser and Peter Mackenzie Smith.

We would also like to extend profound thanks to all the contributors who have given up their time and shared their stories with us.

The Case Studies

INTRODUCTION TO THE CASE STUDIES

Each case study in this volume tells the story of a country that has an economically significant oil and gas sector. We have set out to trace the evolution of the oil and gas industry and, within this, to explore how these countries have addressed the challenges of skills and workforce development. We have considered the political and legal framework for this, the pursuit of workforce nationalisation, the role of local content legislation and the wider role of the local education and training sector. We have also explored what part international partners—both within the industry and across the education spectrum—have played. Our purpose here is to evaluate how successful our focus countries have been in creating a hydrocarbon sector that benefits that country's citizens, that stimulates the wider economy and that builds local education and training capacity.

Every case study has been developed through in-depth interviews with those involved, augmented by rigorous desk research in order to establish the facts. We have, in every case, sought to uncover not only the successes but also the challenges and failures. Where possible, we have included levels of investment, data and analysis of policy changes relating to impact. We have attempted, with each case, to give an entirely accurate reflection of the story and to reflect the contributions of those who took part.

Following each case study, we have included commentary and analysis from the Getenergy Intelligence team—this is designed to give the reader deeper insight into the case and to uncover what we can learn and take away from each case. It should be made clear that the "Getenergy View" represents the opinions of our editorial team and not the inputs of our case contributors (to whom we are eternally grateful).

About Getenergy Intelligence

Getenergy Intelligence was established in 2014 to support the aims of the wider Getenergy group of companies, namely to bring new intelligence to the development of skills and competence across the energy industry. This mission is underpinned by the belief that countries who have abundant natural resources – and the operating and service companies that explore for and produce those resources – have a specific set of responsibilities:

- To grow the energy sector, the energy supply chain and related industries in a way that brings opportunity to citizens and that enables those citizens to play an active role in this economic and industrial growth
- To achieve this through helping citizens to develop the knowledge, competencies and behaviours that are demanded by employers and that will be transferable to other sectors
- To create this opportunity for citizens by building a comprehensive, sustainable and accessible education and training system in partnership with industry
- To build this system locally and to involve local education and training providers in a partnership of equals
- To have a long-term vision around workforce and skills development that sees beyond the immediate skills needs of the energy sector and envisions a future economy that is diverse, dynamic and competitive

Getenergy Intelligence is committed to working with governments, educators, employers and others to create debate, stimulate engagement and, through this process, capture the stories, ideas, opinions, data and case studies that emerge from these activities. The desire to capture, distil and share global good practice is made real through the publication of the Getenergy Guides series and reflects our mission to support and promote the development of effective education and training in hydrocarbon-producing countries.

Case Study 1

Mexico: The Mixed Blessing of Oil Nationalism

The authors would like to thanks David Shields, Director General, Energia a Debate for his help with this case study. We would also like to express our gratitude to Colin Stabler, Consulting Geologist and to Carlos Reyes Abreu, Director General de Energía, Secretaria de Energía, Recursos Naturales y Protección Ambiental, Gobierno del Estado de Tabasco for their time and wisdom.

INTRODUCTION

On March 18, 1938, Mexico's president, Lázaro Cárdenas, announced the expropriation decree, effectively nationalising all the assets of international oil companies operating in the country. This marked the beginning of 70 years of state control over the oil and gas sector in Mexico. In 2012, Mexico began a process of constitutional change that would allow the re-entry of international competition into the Mexican hydrocarbon sector, enabling international oil companies to bid for exploration and production contracts. The reforms themselves, very much the personal project of President Enrique Peña Nieto, will redraw the map of the Mexican energy sector and are driven by the belief that the future of this sector is of profound significance to the health of the economy

Education and Training for the Oil and Gas Industry: The Evolution of Four Energy Nations.
http://dx.doi.org/10.1016/B978-0-12-800974-1.00001-5

and the wellbeing of Mexico's 120 million citizens. The world is now watching to see whether these reforms can boost the country's oil output and bring new opportunities for employment and economic development.

The story of Mexico's oil and gas industry is one rooted in a deep sense of ownership amongst the people, a belief that natural resources are national assets that should be owned and controlled by the people for the people. The concept of local content was, for many years, superseded by the realities of an industry entirely dominated by the state-owned national oil company, Petróleos Mexicanos (or Pemex as it is commonly known), a company established in 1938 and that grew to become one of the largest oil producers in the world. To this day, the vast majority Pemex's employees are Mexican meaning that, since the 1930s, the ability of the Mexican state to educate, train and develop its citizens for the industry underpinned (or perhaps undermined) the strength of the industry as a whole.

In exploring the case of Mexico's evolution as an energy nation, we will examine and evaluate the advantages and dangers inherent in state domination of the sector. We will also explore the reality that lies beyond the visible nationalism that defined the industry for so many years, a reality that reveals an industry much more dependent on foreign expertise that one might think. And we will reflect on the genesis of the new reforms in Mexico (which relate very strongly to workforce challenges) and what these reforms are likely to mean for the industry in Mexico and for the development of a new cadre of skilled people across the country.

© *Shutterstock*

COUNTRY FACTS

- Mexico's population estimate for 2013 was 118,395,054, which is a 1.77% increase from 2010.
- Gross domestic product for 2014 was estimated to be US$2.143 trillion total. Mexico has the 4th largest nominal GDP and the 10th largest in terms of purchasing power parity.[1]
- Mexico has a young population, with 58.5% of the total population aged between 15 and 54 (18.1% aged 15–24 and 40.4% aged 25–54).[2] The age profile of Mexico's population has helped contribute to its economic success.
- The oil and gas industry generates 10% of Mexico's export revenues; total export revenues in 2012 were estimated to be in the region of US$370.9 billion. However, more recent analysis of the state of Mexico's oil industry reveals a sharp decline in the overall amount that the hydrocarbon sector contributes to export revenues – crude exports fell from a peak of over US$49 billion in 2011 to around $36.9 billion in 2014.[3] This is primarily down to a reduction in the amount of oil exported to the USA as a result of increased domestic production there.
- Pemex's oil revenue is the single biggest contributor to the Mexican treasury, supplying roughly one-third of the national budget.
- Mexico is the sixth largest producer of oil globally. However, recent years have seen a steady drop in production levels with estimates showing oil output to be 2.5 million b/d[4] in 2014. This has had a significant effect on Mexico's export revenues and is a factor underpinning the reforms to the sector that are being introduced.
- The decline in Mexico's oil output is related to two key factors. Firstly Mexico's developed fields are maturing. 25% of the country's oil production originates from fields that are 65% depleted. Further to this another 45% of production comes from fields that only contain 25% of their original reserves. Together this means that about 70% of Mexico's oil output relies on mature fields unable to maintain peak rates of production.[5] Secondly Pemex, the national oil company, is limited in its ability to develop untapped fields/reserves due to financial restrictions and a lack of technology and know-how.
- The Cantarell Complex – located 80 km offshore in the Bay of Campeche and made up of four fields – is the largest oil field in Mexico. However, the field is now an ageing supergiant and no longer produces oil at historic rates.

1. "Mexico". *International Monetary Fund*. Retrieved November 2, 2014.
2. Indexmundi – http://www.indexmundi.com/mexico/age_structure.html.
3. Pemex – *Value of Crude Oil Exports, Energy Aspects Calculations*.
4. "Awaiting the Mexican Wave: Challenges to Energy Reforms and Raising Oil Output", *The Oxford Institute for Energy Studies* (June, 2014).
5. Some figures from "Awaiting the Mexican Wave: Challenges to Energy Reforms and Raising Oil Output", *The Oxford Institute for Energy Studies* (June, 2014).

Production peaked, with the assistance of nitrogen injection enhanced oil recovery (EOR) techniques, in 2003 resulting in 2.1 million b/d. However, since 2012, this has sharply decreased and production rates are now below 0.3 million b/d.[6] Estimates show that only a fifth of Cantarell's original 35 billion barrels of reserves remains. The KMZ field, which has, since 2009, overtaken the adjacent Cantarell field in terms of production levels, produced 0.87 million b/d in 2013 and accounts for one-third of Mexico's oil production. The same EOR techniques have been employed to increase flow and production rates. KMZ is now showing the first signs of maturing, with annual production marginally falling.

- Mexico has a mixed economy with industry accounting for 36.6%, services 59.8% and agriculture 3.6%.[7] The country exports cars, electronics and electrical goods, silver, fruit and vegetables, coffee and cotton (as well as oil). These sectors employ a labour force of 78.2 million, of which 33.4% work in industry, 10.7% in agriculture and 55.9% in the services sector.

- Despite Mexico's large and diverse economy – and the wealth that continues to flow from the production and export of natural resources – the country faces serious socioeconomic challenges. 52.3% of the population live below the poverty line[8] and official unemployment stands at 4.5% (2014).[9] These figures raise some serious concerns for a country with aspirations to becoming one of the world's leading economies.

- Corruption is seen as a major issue affecting enterprises and government (including Pemex) with the impact being seen in an inability to further grow and diversify the economy and an increase in the cost of consumer goods and services.[10]

- Pemex, the national oil company, has lost millions of dollars in revenue due to corruption at the state and corporate levels. Mexico ranks 39th on the World Bank Group's Doing Business index, 4 places up from 2014s rank.[11]

- Mexico's workforce is rated by the OECD and the World Trade Organisation as the hardest working in the world when judged by the amount of hours worked annually,[12] yet pay rates remain low and national debt is relatively high, accounting for 37.7% of GDP in 2013.[13] However, the country has made significant strides towards educating and improving the skills, expertise and human capacity of its population.

---

6. "Awaiting the Mexican Wave: Challenges to Energy Reforms and Raising Oil Output", *The Oxford Institute for Energy Studies* (June, 2014).

7. *CIA World Factbook* (2013 est).

8. "Population Below Poverty Line". *The World Factbook*.

9. *CIA World Factbook*.

10. Lilia González, "Sector patronal urge a combatir la corrupción", *El Economista* (April 4, 2011). Retrieved April 27, 2012.

11. http://www.doingbusiness.org/data/exploreeconomies/mexico.

12. Booth, William, "Siesta? What siesta? Mexicans work longest hours in world", *The Washington Post* (May 3, 2011).

13. *CIA World Factbook*.

- Workforce diversification is improving, with the national oil company making important moves to attract more female employees. In 2010, 139,143 employees were male (75.6%) and 44,947 were female (24.4%).[14] Pemex has also demonstrated a commitment to investing in the education and development of its employees with the company spending US$672.6 million in training in 2010. A total of 13,206 courses equalling 736,493 hours of training were undertaken. There were 175,876 employees who participated in courses.[15]
- The U.S. Energy Information Agency forecasts that Mexico's oil production will decline by 600,000 barrels a day by 2020, at which point the country would become a net oil importer.

Gulf of Mexico oil rig in stormy seas.

A HISTORY OF MEXICO'S OIL AND GAS SECTOR

The identification of Panuco-Ebano and Faja de Oro fields in 1910 marked the dawn of Mexico's oil industry. Exploration activities were subsequently taken over by a number of US companies who, by 1911, had begun production and were exporting Mexican oil overseas. At this time, Mexico's oil industry was dominated by the Mexican Eagle Company (a subsidiary of Royal Dutch/Shell). The Mexican Eagle Company accounted for over 60% of Mexican oil production. A number of other American companies were also involved in hydrocarbon extraction, including Jersey Standard and Standard Oil Company of California (now Chevron).

The Constitution of 1917 asserted Mexico's rights and ownership of its subsoil resources. In addition to this, the constitution established the principle of

14. Official Pemex Website – http://www.Pemex.com/informes/social_responsibility/operations/people.html.

15. Official Pemex Website – http://www.Pemex.com/informes/social_responsibility/operations/people.html.

universal access to education as mandatory for all citizens of Mexico. Article 123 of the constitution laid out labour rights, stating 8 hour work days were to be enforced, asserting the right to strike, the right to one day of rest a week and the right to compensation for unfair dismissal from one's position. Article 123 was a critical element of the 1917 Constitution that was designed to alleviate the pressures under which the working classes were living. At the time abuse, hardship and inequality were common and the new constitution established a strong basis for the rights of workers and the responsibilities of employers. The legislation had its roots in the Mexican revolution (1910–1916) during which issues around land ownership and social inequality had risen to prominence.

The 1917 Constitution, which sought to win the sympathies of Marxists, Socialists and gain popular support throughout the country for the then President Carranza, introduced a number of measures which were hugely effective in ending exploitative practices. The Constitution also made reference to rights in relation to Mexico's natural resources. Article 27 stated that:

> *The State will also regulate the exploitation of natural resources based on social benefits and the equal distribution of wealth. The state is also responsible for conservation and ecological considerations. All natural resources in national territory are property of the nation, and private exploitation may only be carried out through concessions.*

Although not fully realised until later reforms in 1938, this set the parameters for the oil industry in Mexico and created the vision for a state-owned sector run by and for the people.

In the immediate aftermath of the 1917 Constitution, the government allowed private foreign oil companies to develop hydrocarbon reserves. The involvement of international oil companies allowed Mexico to become the world's second largest producer of oil throughout the 1920s. However, political difficulties began to emerge during this time as it became clear that revenues from the sector were increasingly flowing out of the country. These issues were exacerbated by the global oil glut in the late 1920s and early 1930s that sent prices plummeting and saw a steep decline in oil revenues. This challenging economic context underpinned rising disaffection with the way international operators conducted their business in Mexico. Popular resentment was further fuelled by the fact that international companies paid nationals around half of what overseas employees were paid.

Within this context, it became increasingly likely that the Mexican government would assert its rights over hydrocarbon reserves and make good on the intentions of the 1917 Constitution. Tensions eased in 1928 when the Calles–Morrow agreement reaffirmed the rights of international operators in Mexico. However, after years of mounting pressure a turning point came; a strike by oil workers in 1937 forced Mexico's government to take action. Mexican nationals wanted increased pay and social services. They protested at what they perceived as exploitation by the oil companies who had, for years, paid the local workforce poorly whilst investing in foreign talent. Recognising the need to have foreign companies participate in production, President Lázaro Cárdenas initiated a settlement whereby a new labour agreement would seek to address the frustrations and concerns of native Mexicans working in the oil sector. However, the international operators failed to respond positively to this new agreement with many taking a 'business as usual' approach. In response, and in the face of mounting political and social pressure, President Cárdenas announced the expropriation decree on March 18, 1938 effectively seizing the assets of international oil companies and nationalising Mexico's oil and gas sector. In the same year, Pemex (Petróleos Mexicanos) was created as the national oil company of Mexico and took control of all exploration and production activities.

The consequences of this radical shift in policy were immediate and significant. International oil companies responded to what they perceived as unwarranted hostility from the Mexican government by placing an embargo on Mexican oil, resulting in a drop in oil exports of around 50%.[16] Diplomatic relations between Mexico and Britain were damaged and Germany became the principal buyer of Mexican oil. Reactions in the USA were mixed, with President Roosevelt recognising the need to carefully manage relations with their near neighbour. Despite pressure from some elements within the US government, the administration chose to maintain relations with the Mexican government, leaving the US oil companies who had been active in Mexico to take up the fight for what they saw as unfair treatment. Against the backdrop of World War II – and increasing pressure from the US government – demands for compensation were finally settled in 1942 when the US and Mexican governments signed the Cooke–Zevada agreement. Under the agreement, Mexico would pay US$29 million in compensation to a number of American-based oil companies. Subsequently, British oil companies received a $130 million settlement in 1947. This marked the end of a turbulent chapter in the history of Mexico's energy industry and the beginning of a long period during which international oil companies would play little part in Mexico's oil and gas production and exploration activities. Although attempts were made by international oil companies to re-enter the Mexican oil industry, the increasing strength and dominance of Pemex created too many barriers for international companies who, faced with opportunities in other parts of the world, took their business – and expertise – elsewhere.

The impact of this period in the history of Mexico's oil and gas sector has been profound. The history behind the nationalisation of the Mexican oil industry has been taught to generations of Mexican school children and is framed as a story of great national pride and liberation from American imperialism. As such, the growth of Mexico's oil nationalism – and the emergence through the second half of the twentieth century of Pemex as a force within an international oil and gas landscape dominated by some of the world's most powerful companies – has become a narrative that's entwined with the political and economic fortunes of the country as a whole.

---

16. U.S. Department of State, Office of the Historian.

The radical reforms of the Mexican oil industry that began in 2012 are directly linked to the twin challenges of falling production levels and an inability to capitalise fully on remaining reserves. Essentially, the Mexican government recognised that without the investment, expertise and technology of the international sector, the country would be unable to replenish its reserves. Independent studies have suggested that capital expenditure in the region of USD $830 billion is required in order for Mexico to fully develop and exploit remaining oil and gas deposits. That figure would appear to be way beyond the means of Pemex, who reported revenues of around USD $120 billion in 2013 (and even more so if one takes account of the funnelling of revenues into government coffers via taxation). Pemex – and, by extension, the Mexican oil industry – has survived and, in some senses, thrived by controlling the industry, focussing on the production of a small number of enormous fields and building a deep understanding of how to get the most from what was easily available to them. As the global hydrocarbon industry moves into a new phase – one where unconventional methods of exploration and production will become the norm and cost and complexity increase – the failure to create a long-term plan for technology, investment and skills has cost the country dear. The days of Mexico's oil industry as a signifier of the independence and strength of a great nation are numbered. A new future looms large.

A HISTORY OF EDUCATION, TRAINING AND WORKFORCE DEVELOPMENT

The history of education and training in Mexico during the twentieth century is deeply connected to events that have also shaped the evolution of the oil and gas industry. The Mexican Revolution – which had its genesis in a struggle for basic human rights – resulted in constitutionally guaranteed educational and social benefits for all. The Constitution of 1917 had as one of its key pillars access to a secular education. The Constitution also gave powers to the federal government over education, including control over structure and curricula. In 1921, the federal Secretariat of Public Education was created. At this time, the nationalism that was to shape the oil industry a few years later became entrenched into the public school system in Mexico, something that remained in place until 2001.[17]

The advent of World War II provided the impetus for a more focused approach to workforce and skills development. As the United States looked to Mexico for human resources, the government undertook measures to prepare its workers for industry and created the *Camara Nacional de la Industria de Transformación* (now known as CANACINTRA and still operational today). By the end of the war, around 300,000 Mexicans had worked in 25 of the United States, creating a precedent for the illegal migrant-worker market in North America.[18] Mexico's industrialisation during the 1950s saw a continued focus within the Mexican

---

17. http://education.stateuniversity.com/pages/979/Mexico-HISTORY-BACKGROUND.html.
18. Ibid.

education system on providing adequate training for the new industrial workers with political leaders realising that industrial growth would only happen if the country was able to build a workforce with the right skills. Alongside this was a drive to expand and improve higher education. The construction of the new University City built to house the National Autonomous University of Mexico was completed in 1952 with the site spreading over 3 square miles. At the same time, an intellectual movement emerged and between 1940 and 1951, the *El Colegio de Mexico,* the *Escuela National de Antropologia e Historia* and the *Instituto de Historia of the National University* were established, in part to create a knowledge base that could define a new understanding of Mexican history.

The signing of the North American Free Trade Agreement (NAFTA) in 1994 created a fresh set of challenges for Mexico's educational system. Public education provision at primary and secondary level was not up to the standard of Mexico's new trading partners, Canada and the US. The agreement also restricted the ability of Mexican children living in border areas to go to US schools. The subsequent collapse of the Mexican economy following the signing of NAFTA forced an end to the subsidized university system where tuition fees had remained at the same level since 1948. Overnight, these fees went from a few cents per term to the equivalent of $70.00 leading to massive student protests at the National Autonomous University.[19]

Today, the education system in Mexico is run by the Secretariat of Public Education (SEP). However, 85% of primary and secondary schools are controlled by individual states.[20] The SEP is the best funded department of the Mexican government with spending on education set at a minimum of 8% of GDP (one of the highest figures in the world). Primary, preprimary and lower secondary schools provide basic education and account for 78% of students overall. 62% of students in basic education attend primary school, 23% lower secondary and 15% preprimary.[21] Upper secondary school allows students to follow one of three specialisms:

- General secondary (which accounts for 60% of enrolment) provides an academic education designed to prepare the student for university.
- Technological secondary (which accounts for 29% of enrolment) offers students preparation for entry onto technological degree courses or provides students with a technical qualification for employment.
- Technical professional secondary (accounting for the remaining 11% of enrolment) provides students with a technical vocational education and leads to qualifications that enable students to enter the labour market.

Although the education system has historically struggled to achieve high rates of literacy – and enrolments into secondary education have been low in comparison to other developed nations – the situation has been steadily improving with rapid increases in the number of available secondary places mirrored by an increase in enrolments.

The tertiary system in Mexico presents a mixed story. In 2004, only 2.35% of the population were engaged in any kind of tertiary studies[22] (representing a worryingly low figure, particularly in a country with such a young population). In 2002, there were 1550 institutions of higher education, around 600 of which were public. The tertiary system is delineated as follows:

- 52 Federal and State universities
- 189 Technological institutions
- 54 Technological universities
- 863 Private institutions (including universities, institutes and centres)
- 392 Teacher training colleges

The perception of the tertiary sector in Mexico is that the university system generally offers a very high standard of academic education to students. Considerable funding is channelled into university education by the SEP and the country can boast a number of world-class academic institutions. However, the technical and vocational colleges (including the technological universities) have a less favourable reputation.

A 2009 report by the OECD into Mexico's system of technical and vocational education found a number of profound and systemic issues that needed addressing. These included the following[23]:

- A lack of effective coordination and coherence within upper secondary vocational education and training
- Poor linkages between the vocational education and training system and employers, illustrated by the low level of involvement of employers in VET policy development
- Vocational education and training qualifications that are not regularly updated and have limited recognition in the labour market
- VET teachers and trainers that have not received sufficient training in teaching methods and approaches
- A wide variation both in the quantity and quality of workplace training for technical and vocational students
- Weaknesses in the availability and use of data for policy making purposes and to inform stakeholders

As we look at the challenges of workforce and skills development for the oil and gas sector, we can see that Mexico offers a complex picture in relation to education and training. On the one hand, the country has seen steady industrial growth since the 1950s, significant investment in education at every level, some

22. Ibid.
23. *Vocational Education and Training in Mexico Strengths, Challenges and Recommendations* (OECD, 2009).

success with efforts to build a skilled workforce and real quality within the university system. However, education beyond secondary level remains accessible to only a minority of students, a preference for academic study prevails and quality and relevance across the technical and vocational system remains questionable. As the engine of the Mexican economy, does the oil and gas sector reflect these realities or is there a different story emerging?

THE STORY OF WORKFORCE AND SKILLS DEVELOPMENT FOR THE OIL AND GAS INDUSTRY

In considering the effectiveness of Mexico's workforce and skills development efforts, it is, perhaps, instructive to reflect on the recent oil reforms that are reshaping the oil and gas sector and that will change the landscape in Mexico for a generation. Put simply, the shift away from oil nationalism and towards a more international and competitive market is entirely pragmatic. When plans for the opening up of Mexico's energy sector began to emerge in the late 2000s, opposition was fierce. The unique place that Mexico's oil industry has in the national psyche meant that the political risk associated with any denationalisation project was high. However, the political leadership recognised the cold realities of an industry that had failed to innovate, had underinvested and been wasteful. Without direct foreign investment and, crucially, foreign skills and expertise, the Mexican oil and gas industry would continue to falter and would not be able to capitalise on remaining reserves. In short, Mexico could no longer do it alone. The subsequent dissolving of opposition to the denationalisation plans is evidence of the wider acceptance of this reality. And although money was, and continues to be, a key issue for the Mexican energy sector, people and skills are also at the heart of the challenge. It is within this context that we can consider the history of Mexico's efforts to build and maintain a workforce for the oil and gas industry.

Pemex gas station in Tijuana, Mexico. *(Joseph Sohm/Shutterstock.com)*

Pemex and the Impact of Oil Nationalism

Pemex is the dominant force in the oil and gas industry in Mexico. Since its establishment in 1937, it has grown to become Mexico's most important organisation in terms of contribution to the economy. By one simple measure, we might consider that Pemex has been extraordinarily successful in building a workforce – the statistics speak for themselves:

- As of 2010, Pemex employed a total of 184,090 people (118,749 in permanent positions and 65,341 on temporary contracts)[24]
- In the same year, Pemex accounted for US$27 billion in total wages and benefits[25]
- Of those employed, 63,766 were engaged in exploration and production activities; 58,215 were engaged in refining; 13,837 were engaged in gas and basic petrochemicals and 17,066 in petrochemicals[26]
- In 2010, Pemex invested US$672.6 million in training encompassing a total of 13,206 courses and 736,493 hours of training with 175,876 employees involved in some form of training[27]

The vast majority of Pemex employees are Mexican meaning that Mexico has achieved what many other oil- and gas-producing nations have sought to do: build an industry that is led, managed and operated by nationals within the structures of a successful national oil company. However, workforce nationalisation is not an end in itself. A thriving oil and gas sector relies on a wide range of skills, experience, knowledge and competency. The gradual diminishing of Mexico's effectiveness in both production and exploration (particularly in relation to the comparative flourishing of neighbouring Brazil) can be traced to the failures of Pemex to adequately address skills and workforce development (just as Brazil's successes are a measure of the dynamism and vision of the national oil company Petrobras). But before we explore these failures in more detail, let us recognise the progress that Mexico and, by definition, Pemex have made over recent decades.

By establishing a national oil company with sole responsibility for the industry – and by ensuring that company was mandated to employ nationals – the government of Mexico created a context within which the only option was to train, educate and develop Mexicans. Although some international expertise was retained within Pemex in the early days of oil nationalism, the focus was to develop talent within the country and establish Pemex as the centre of gravity for the industry and for the development of talent. In terms of the efficiency of Pemex in this regard, it is not easy to ascertain. Through successive administrations, politicians have held up Pemex as a model of industrial

---

24. Figures supplied by Pemex.
25. Official Pemex Website – http://www.Pemex.com/informes/social_responsibility/operations/people.html.
26. Ibid.
27. Ibid.

success whilst critics spoke of mismanagement and corruption. What is undeniable is that Pemex steadily grew its activities, production rose, the organisation expanded and budgets soared.[28] A number of observers also recognise that, over the course of many decades leading the industry, Pemex has developed particular expertise within specific technical fields, allowing the company to maintain comparatively high levels of production without becoming overly reliant on foreign talent. Working in partnership with universities – and through a longer-term process of building the in-house experience of committed staff – Pemex boasts considerable talent within the fields of oil engineering and oil geoscience. Although Mexican oil fields are characterised as large and easy to exploit with a low cost of production in accessible offshore locations, they do have their own specific geological characteristics that are unique to Mexico. This is particularly true of the oilfields in the Tabasco area. Pemex has developed – and, crucially, retained – the expertise and knowledge to design projects for these fields. What's more, the strength of the university system, and the ongoing funding it receives (more of which below), means that Pemex has a continuous pool of talented engineers to recruit from. There is also a broad recognition that Mexico boasts a good level of technical expertise in relation to the building of offshore platforms.

In addition to this, Pemex has had close ties with the Instituto Mexicano del Petróleo (Mexican Institute of Petroleum or 'IMP' as it is commonly known) since its inception in 1965. In fact, the establishment of IMP is bound up in the history of Pemex's workforce development vision with IMP describing their creation as being 'born by the initiative of the then Pemex's general director, Jesus Reyes Heroles, who recognized that planning and development of the petroleum industry have to be congruent with the needs of a mixed economy. Because of this, he considered that it was necessary to promote petroleum research and form human resources that could boost their own technology development'.[29] In the years that followed, IMP grew to become a key part of Pemex's strategy for skills and workforce development. The Institute provides dedicated training and research to the Mexican oil and gas industry and has 300 staff involved directly in the provision of training and 3000 employees engaged in research and development activities.[30] IMP was, until 1990, funded directly by Pemex. However, the relationship between IMP and Pemex was subsequently redrawn and the Institute now negotiates contracts with Pemex for the provision of training and research. IMP concentrates mainly on upstream training and specifically on drilling. In 2004, their annual training budget from Pemex apportioned around £22 million to upstream training courses, £5 million to downstream and £1 million to gas.[31] The model

28. Robert J. Shafer and Donald J. Mabry, *Neighbors—Mexico and the United States: Wetbacks and Oil* (Chicago: Nelson-Hall, 1981).
29. Instituto Mexicano del Petróleo Website – http://www.imp.mx/acerca/en.php?imp=historia
30. *Mexico: Training and Education Opportunities* (UKTI, 2004).
31. Ibid.

that Pemex developed with IMP was to provide the Institute with an annual needs assessment report and then collaborate on the design and development of bespoke courses based on this assessment. This required IMP to be highly responsive to Pemex's skills needs and ensured that the partnership lay at the heart of Pemex's workforce development activities. However, in recent years, Pemex has shifted strategy towards a more competency-based approach and IMP has seen their proportion of Pemex's training business diminishing (with the practical training offered at Mexico's technical universities winning market share from IMP).

There is no doubt that the nationalising of the Mexican oil industry – and the creation of Pemex as the vehicle for all exploration and production activities – simplified the task of workforce and skills development. With only one company operating on any meaningful level, and significant political and economic backing for education and training initiatives, Pemex was able to design and build a workforce to their particular requirements. And whilst there has clearly been some success emerging from this scenario, there have been a number of challenges that, in part, led to the ultimate erosion of Pemex's efficiency and have ultimately diminished the ability of the country to maintain control over the sector:

The Myth of the 'Mexican' Energy Sector

One of the key issues with the development of the Mexican oil industry – and a factor behind the skills and talent shortages that have been endemic for some years – is the fact that although none of the international oil companies were permitted to operate in Mexico for decades, the international service companies grew increasingly important, particularly in the early part of the twenty-first century. Pemex continued to claim that their industry was wholly Mexican whilst contracting out to a growing band of overseas providers who brought the skills and expertise that Pemex was unwilling or unable to develop. This pattern developed to a point where Pemex became heavily reliant on international service companies to complete critical exploration and production projects. As a consequence, Pemex has not been driven to build in-house or in-country expertise – particularly in the more technical disciplines related to unconventional hydrocarbons – but rather has developed a culture of subcontracting. Not only has this hindered skills development across the industry but it has also meant that many Pemex employees have ended up being contract managers rather than engaging in the more technical aspects of the job for which they were trained. Today, it is well known that the more technologically demanding projects are, the more likely they will be completed by international companies (with the simpler contracts undertaken by Mexican service companies or by Pemex themselves). Although a majority of available contracts across the sector are awarded to Mexican companies, the largest contracts are won by international providers – this reflects the inability of native Mexican companies to compete for technically challenging projects.

The Culture of Pemex

Over many years of operation, Pemex became a large and, some would say, bureaucratic organisation that suffered from some of the worst characteristics of a state-run enterprise. Operating within the context of a highly unionised workforce, those who were employed by Pemex knew that they had a job for life. This did not mean that the organisation lacked self-motivated, ambitious individuals who wanted to progress but there was no culture of ambition. Without competition within the organisation – and without the influence of different cultures – too many Pemex employees 'took the paycheque' without really engaging fully with the job. In addition, Pemex predominantly recruited Mexicans and, as a result, became very insular. The company reflected the common perception amongst ordinary Mexicans that foreigners would only come to Mexico to make money. International investors made their way into other areas of the economy successfully, but this led to resentment, especially in the banking sector. With a majority of banks in Mexico owned by Spaniards, Americans or Canadians, Mexicans became wary of the motives of foreign companies. Pemex reflected a lack of openness to the influences and contributions of the outside world and fell behind as a result. What's more, the company became synonymous with what one commentator describes as 'notorious inefficiency'. The same commentator goes on to describe the company as 'laden with bureaucracy, teeming with superfluous workers, and... thwarted by corruption'.[32] This does not, in itself, reflect any inherent issue with a dominant state-owned company running the energy industry but, in the case of Pemex, it does suggest a failure to tackle a working culture that had a profoundly negative effect on productivity.

The Wrong Approach to Recruitment

Issues around recruitment had a significant role to play in the development of Pemex's workforce. For many years, recruitment was not based on merit but rather was driven by two factors: first, there was a widely acknowledged culture of nepotism with many senior positions taken by those who were well connected within government and within Pemex itself. Second, the dominance and power of the unions meant that many other jobs – particularly skilled jobs – were given to those who had friends or connections in the labour movement. This meant that, at the recruitment stage, the quality or ability of candidates became a secondary consideration. It is inevitable that this was to have a damaging effect on workforce development within Pemex. Without the right basic skills, employees have a far greater distance to travel in order to become effective, competent contributors to the business. Within this context, the notion of workforce planning – and the connection this should have with education and training – crumbles.

---

32. Fortune Online: http://fortune.com/2014/08/14/Pemex-oil-black-gold/

The Lack of a Plan

Although there is evidence that, in recent years, Pemex has taken steps to be more systematic in its approach to education and training, the history of the company in this regard is one of little planning and of delegating education and training decisions to individual managers. It is certainly not the case that training opportunities were not available to Pemex staff. Rather the issue has been that the nature of the training you received would be defined by the individual philosophy of your manager. Without a coherent plan that operated across the whole organisation – and without a competency framework to sit around it – the challenge of continuing professional development cannot be met.

Loss of Technical Talent to Management Roles

The lack of coherent planning within Pemex had an additional negative affect on the business. Over time, the company has invested in higher level skills, particularly in relation to geologists, geoscientists and petroleum engineers. Many have been sent abroad to complete Master or Doctorate studies and have then returned, bringing with them a broad academic and technical expertise. However, in too many cases, these individuals became managers of people and projects rather than being deployed where their talents would have most impact. This also meant that there was little by way of transfer of knowledge between those well-educated employees and those who were looking to learn on the job.

Failure to Adapt

Many of the other issues outlined here combine to reflect a core weakness in Pemex's approach to workforce development – a failure to adapt to the changing nature of the Mexican industry and the evolving demands of the sector. Ultimately, it has been this failure above all others that has contributed to the country needing to reach out for international expertise in order to bolster exploration and production. There is a strong sense that Pemex became complacent, particularly in relation to building the technical competencies that would be needed to sustain the business as the productivity of mature fields diminished. Any successful education and training system needs to be responsive to shifting industry dynamics but the evidence suggests that Pemex did too little in this regard.

Ultimately, it is difficult to judge the impact that Pemex's workforce and skills development activities have had on the industry as a whole and on the relative employability of Mexican nationals. What seems clear is that the commitment to building an industry that was led by Mexicans brought with it certain benefits – this approach created unique opportunities for Mexican nationals with the right skills, the right mentality and the capacity to learn. However, it seems that Pemex was dogged by short-term thinking, lacked a vision for education and training and failed to invest during periods of plenty. As a result, the oil and gas workforce in Mexico has needed the support of international service companies for sometime and, more recently, has recognised the shortfall in talent and the impact this is having on the longer-term viability of exploration and production activities.

Tertiary Education and the Oil and Gas Industry

As mentioned above, the Mexican higher education sector has, over many years, played a part in the development of highly skilled employees for the oil and gas sector. There is a long history of petroleum engineering, geology and geoscience within Mexico and much of this talent has stayed within the country. The university sector has subsequently benefitted from ex-Pemex employees going back into academia (particularly those who have retired). That said, the availability of courses in oil- and gas-specific fields has been an issue, with strict controls on the institutions able to award degrees in these fields. As an example, the first petroleum engineering degree in the Tabasco region (one of the key oil-producing regions in Mexico) was launched in the late 2000s (very late given how long the region has been actively producing hydrocarbons). There is now more of an open market in the education sector, particularly within the university system (which is a mix of public and private provision) but the sector has arguably been slow to adapt to the new demands of the industry and is having to catch up as a result.

Alongside the successes claimed by the university sector in Mexico, there are indications that their contribution fell short. First, there was a clear bias during the later part of the twentieth century towards sending the brightest and best candidates overseas (rather than educating them in Mexico). Even though many of these individuals came back to take up positions with Pemex, the fact that they completed their studies overseas is a mark of the lack of faith in the higher education system in Mexico in comparison to what the best of the international sector could offer. Further evidence supports the view that those individuals graduating from Mexican universities were often in need of considerable retraining in order to make them competent to do the job (although one might observe that this is a well-worn argument in most countries around the world). One of the challenges facing universities (and common to the vocational sector as well) was that many candidates coming from the Mexican school system lacked the basic functional skills needed for more technical types of studies (with numeracy a particular problem).

Outside of what one might term 'academic study', Mexico implemented a hybrid university model aimed at addressing skills development. The Mexican technological universities (or UTs) system was established in 1991 after a detailed analysis of the approach to vocational education in UK, Germany and Japan and drew on different elements from each country. In 2002, there were 54 UTs throughout Mexico, awarding qualifications up to Higher National Diploma (HND) level in a variety of disciplines.[33] The subjects available for

---

33. *Mexico: Training and Education Opportunities* (UKTI, 2004).

study at the UTs are focused on industry needs both in terms of content and structure. Courses are designed to develop job-specific skills with around one-third of classes theory-based and two-thirds purely practical.[34] The programmes offered are intensive and involve around 3000 hours of study over a 2-year period. Courses conclude with a 4-month industrial internship. The focus of course content is also influenced by the region in which programmes are offered. Two of the technological universities specialise in oil and gas subjects with both located in the South of Mexico – the Universidad Tecnológica de Tabasco in Villahermosa (UTT) and the Universidad Tecnológica de Campeche (UTC). The UT system is overseen and administered by the SEP (Secretariat of Public Education) and was established to ensure that higher education was producing the right kind of graduates for industry. However there has been ongoing issues around the disconnect between workforce requirements – the type of skills needed and the number of people needed with those skills – and what the technical universities actually deliver. That said, the technical universities are now looking to implement a competency-based system (bringing them in line with other vocational institutions in other parts of the world) and are beginning to compete for Pemex training contracts that previously went to the Mexican Institute of Petroleum.

Amongst the other institutions of note within the Mexican tertiary system who play a role in the education and training of nationals for the oil and gas sector, Universidad Autónoma Del Carmen (UNACAR) is located in Ciudad Del Carmen and has been successful in building a strong relationship with Pemex around the awarding of degrees in oil- and gas-related subjects. UNAM (the National Autonomous University of Mexico) is the oldest university in Mexico and is involved in a wide range of university-level research. UNAM has an agreement with Pemex to deliver a range of postgraduate courses in exploration and production. It will be interesting to observe how these institutions – alongside the private education providers – respond to the influx of international companies into Mexico. Relationships with Pemex will undoubtedly change and it will be the institutions who are able to adapt to the new industrial landscape who will thrive.

The Role of the Private Sector

There has historically been very limited private training provision within the oil and gas industry in Mexico. This is largely due to the dominance of publicly funded bodies like IMP and a preference for utilising technical universities and a select few higher education institutions. There are some private training providers operating in the south (Villahermosa and Ciudad del Carmen) where oil production is concentrated but companies have faced real challenges around accessing the market with Pemex contracts particularly difficult to secure within an environment where IMP and other publicly

---

34. Ibid.

funded institutions are not required to tender for contracts (but are, instead, able to draw up agreements directly with Pemex for the provision of education and training services).

That said, there have been isolated but notable contributions from the private sector over recent years (including the work of Schlumberger-owned Next) and the heavy presence in the country of oil service companies (including Schlumberger, Halliburton, Baker Hughes and others) has brought with it a degree of training expertise. With the reform process gathering momentum, the impact on the private training sector is likely to be seismic with opportunities for internationally accredited providers expanding rapidly and a new market for local provision growing out of the arrival of international oil companies. In this regard, the degree to which the government is willing or able to implement and enforce a local content policy could have significant ramifications for the private training sector.

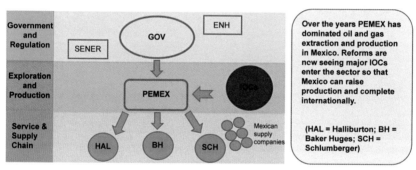

The Oil and Gas Sector in Mexico.

CONCLUSIONS

In assessing the effectiveness of Mexico in building and oil and gas workforce, we have to acknowledge a mixed picture. The Mexican oil and gas sector remains an enormously important part of the economy, generating revenues for the government and employing a large number of Mexican nationals both directly and indirectly. The country can boast a level of expertise born out of a commitment to higher education allied to decades of experience. However, one cannot help wonder what might have been for the Mexican oil and gas industry had the government and, by extension, Pemex developed a clear system of sector skills planning and invested more heavily in building a coherent approach to technical and vocational education that could underpin and support the sector as times changed and industry demands shifted. The sense that Mexico has somewhat been left behind by the international oil and gas community is hard to shake. Failure to build the means to develop the right skills and competencies has contributed to Mexico's current need for reform. Now, the world is watching to see how these reforms might give Mexico a second chance to turn it's abundant natural resources into a lasting legacy of skills.

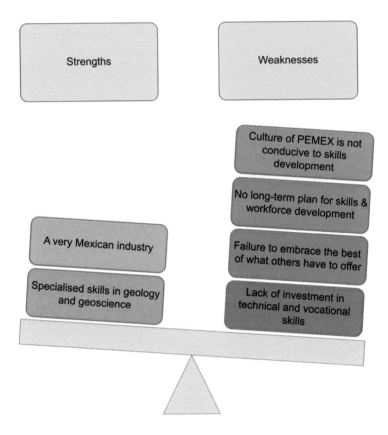

PERSPECTIVES ON THE FUTURE

Reports in 2014 suggested that a number of the available concessions on which international companies were bidding could offer production at an incredibly competitive price (with some estimates putting that price at $20 a barrel). This means that the incentives for international companies to win concessions in the newly opened Mexican oil sector are significant, particularly if oil prices remain low for any length of time. The expansion of exploration and production activities is, therefore, a reality that will bring with it a new demand for skills, expertise, education and training. Furthermore, the role and purpose of Pemex is going to change. What does this all mean for the sector in Mexico and for those engaged in building the new energy workforce that will, it is hoped, transform the country's industry?

- The requirement for a skills 'ecosystem' is now being recognised by many of those engaged in the education and training of Mexicans in-country. Mexico will need to look at models of workforce and skills development from other parts of the world in order to learn and then implement their own approach that is locally fit for purpose. The 'sector skills council' approach is now gaining traction and is likely to be further developed in coming years. This type of system gives industry the power to influence skills and workforce development policy and practice and is what is now needed for the emergent Mexican oil and gas sector. If implemented correctly, such a system will create much better data on which training and education providers can build their offer.

- The influx of IOCs into Mexico will create new jobs and new opportunities for the citizens of Mexico. However, the Mexican public and private education and training sector will need to move fast (and expand) if it is to create the basis for a new approach to skills development that will enable Mexicans to take advantage of these new employment opportunities. If this doesn't happen, the skilled workers of the US, Canada, Venezuela and beyond will be willing and able to take advantage of employment opportunities. Mexico has already absorbed a wave of oil professionals leaving Venezuela in light of the tightening grip the government has over the sector. These professionals quickly found that their expertise was needed in Mexico and were readily absorbed.

- The Mexican government is now having to address the issue of local content (something that was less relevant when Pemex ran the industry). There are signs that some of the international companies interested in bidding for concessions in Mexico have concerns over the implementation of a local content policy. However, an opposing (and perhaps more credible) view is that the focus for the short to medium term will be on increasing production and that the Mexican government is unlikely to put any impediments in the way of international companies keen to get the oil and gas flowing. The longer-term role for local content policy is likely to be a growing area of interest as the sector diversifies.

- The future for education and training will have to involve a much greater degree of partnership and collaboration between international oil companies, local and international education and training providers and, potentially, Pemex (although the role that Pemex will play in this is far from clear). The Mexican education and training market will need to evolve quickly to meet the needs, requirements and standards expected by international operators and, to do this, may well need the assistance of education and training providers from other countries who are able to bolster the in-country offer. Part of this evolution will require investment in infrastructure in order to provide the facilities and equipment now required to train a workforce for the twenty-first century oil and gas business.

- While Mexico offers significant opportunities to international oil companies, ongoing financial and political challenges will limit foreign investment until after 2016, meaning that it will be some years before any real impact will be seen on Mexican production.[35] The challenges facing existing fields alongside

---

35. "Awaiting the Mexican Wave: Challenges to Energy Reforms and Raising Oil Output", *Oxford Institute for Energy Studies.*

beleaguered midstream and downstream sectors will need to be addressed as part of an ongoing reform programme.

- Pemex will be forced into the position of having to compete with other international companies involved in operations. Their staff are not used to this kind of working context and are more familiar with supervising contractors. It will take a re-engineering of Pemex (and considerable programme of retraining) to build a new organisation capable of succeeding in this new market. That said, the right of Pemex to exploration and production activities is protected by government mandates. This issue represents one of the main uncertainties surrounding the industry in Mexico in 2015. Pemex is likely to receive preference over an initial group of fields but unless the company is able to be competitive, it will be relegated to operating the relatively simple fields that it has learnt to exploit well and foreign companies will dominate the harder-to-reach hydrocarbons, from the deepwaters of the Gulf of Mexico to the shale plays near the US border.[36] Though Pemex have been given priority in bidding rounds, so far they have largely opted to retain control over their current fields. There is a definite sense that Pemex is only able to reach the low hanging fruit, but increased competition should hopefully foster a culture of ambition within the company.
- Some commentators suggest that there will simply not be enough Mexicans to fulfil the jobs that will be created as IOCs re-enter the Mexican market. This means that foreign companies will have to train people on the job and bring in expatriates to fulfil roles. Again, the detail around local participation will be critical here in defining the future shape of the workforce.
- Estimates suggest that around 70% of Mexico's proven reserves have already been consumed. This means that there is a growing awareness within Pemex and beyond that the future for the industry will be about secondary oil recovery, Enhanced Oil Recovery (EOR) and about finding new reserves. This is easier said than done and will require a new focus on the development of technologies and technical skills.
- There is some focus on the potential of deepwater oil with IOCs particularly interested in this. However, it is likely to be at least 6 or 7 years before this oil will start to produce and first production may not be very high. These timescales will have an impact on the economy in Mexico but also offer the possibility of a window of time for developing the skills in Mexico that can meet these future demands.
- The challenge of reforming Pemex will be part of Mexico's narrative over coming years. Changing it from being a 'sprawling bureaucracy [and the] poster child for corporate dysfunction'[37] to an organisation able to grow and

---

36. Fortune Online: http://fortune.com/2014/08/14/Pemex-oil-black-gold/.
37. Ibid.

compete will not be easy. The hope is that the current CEO – who is committed to redefining the organisation – will recognise the need to reform not only the culture and structure of the company but also the way it addresses skills and workforce development challenges.

The Getenergy View

- The story of Mexico's emergence as an energy nation tells us much about the dynamic between national oil companies and workforce development. In the case of Pemex, we can see that operating a state monopoly has had its benefits. If the Mexican hydrocarbons sector had bent to the will of the international oil companies, the goal of oil nationalisation would have remained a distant dream. In one sense, the creation of Pemex was a bold move that created opportunities for Mexicans and generated decades of oil revenues for successive governments. Although one can criticise the way in which Pemex has been managed, the company has lead Mexico to the top table in terms of exploration and production activities. By sheer force of numbers, the company has built a workforce that, to a point, is able to maintain the industry at a level that remains vital to the health of the wider economy.
- That said, there have been clear failures and these have become embedded into the company's operations. The working culture at Pemex has had a major impact on skills development. The historically nepotistic approach to recruitment coupled with a lack of accountability and labour laws that are overly favourable to employees have, collectively, created an organisation that is bloated, lacking in dynamism and unable to maintain a high standard of working practices. In short, Pemex has, for many years, not been run like a business but has exhibited the worst characteristics associated with a state-run enterprise.
- In addition to challenges of culture, Pemex has also failed to implement a coherent skills development policy underpinned by a deep understanding of the existing competencies of the workforce. By devolving responsibility for training and development to individual managers, the company created a complex system that relied too much on individual decision-making rather than cohering to a plan and vision. There are indications that the company is moving towards a competency management approach but this should have been in place many years before.
- Mexico has clearly suffered from an affliction common amongst many nations: there has been too much of a focus on higher education – and on the role of the university – and not enough of a focus on technical and vocational education. Although there are undoubted strengths within the higher education system in Mexico, building technical and artisanal skills is a critical part of the industrial strength of any nation. This is particularly true when applied to the oil and gas sector. Whilst there have been efforts in this regard, they have had limited impact and have left Mexico short of the talent needed to see the industry thrive.

The Getenergy View—cont'd

- The monopoly that existed for decades across the hydrocarbons sector has, to a large extent, been mirrored by the provision of oil and gas education and training. Those Mexican institutions able to deliver courses relevant to the industry have been able to win education and training contracts from Pemex without the need to compete with others. Although competition in itself is not the answer to skills development questions, those institutions who have benefitted from Pemex's patronage over the years have had little incentive to innovate in terms of their curricula, pedagogy or approach. This means that the education and training experience falls behind international standards and leaves these institutions vulnerable to the imminent entry of overseas competitors.

- Employing nationals is not an end in itself. The oil and gas industry has, over many years, provided significant employment opportunities for Mexicans (both directly and indirectly). However, for the industry to thrive, those employed need to be properly trained and need to be equipped with the skills and competencies that the industry actually needs. If local participation is not accompanied by a coherent approach to education and training, the industry will struggle.

- The goal of oil nationalism is not unreasonable. Nor is it difficult to understand or justify. However, the failure of the Mexican oil and gas sector to embrace international partnerships has, ultimately, weakened the industry and led to the sell-off of concessions. Although there is no doubt that international participation needs to be carefully managed, it is equally true that in a fast-moving sector like oil and gas, partnerships with global companies can bring a huge amount to the table. The ideal for the new Mexico will be to engage positively with the international community without losing control or ownership of what remains a critical sector of the economy.

Case Study 2

Nigeria: The Story of an Ailing African Oil Giant

Chapter Outline

This case study draws significantly on the work of Dr Jesse Salah Ovadia, Lecturer in International Political Economy at Newcastle University to whom the authors are indebted. The authors would also like to thank Dr Ibilola Amao, Principal Consultant at Lonadek, and Kabir A. Mohammed, Former Executive Secretary, Petroleum Technology Development Fund for their invaluable contributions.

INTRODUCTION

When we talk of the 'resource curse', Nigeria is the country many consider to be the archetype in this regard – a failed oil state that has struggled to achieve the economic and social progress that its abundant natural resources might afford. Despite generating billions in oil revenues, the country has little to show for this wealth, particularly in relation to improvements to education and training. Insufficient investment, corruption and waste have undermined efforts to create a positive and lasting legacy that benefits the wider population. Elements of the political elite have been responsible for misspending or misappropriating resource revenues and have amassed

Education and Training for the Oil and Gas Industry: The Evolution of Four Energy Nations.
http://dx.doi.org/10.1016/B978-0-12-800974-1.00002-7

untold wealth as a result. 85% of the country's oil wealth lies in the hands of 1% of the population (although this is not a situation unique to Nigeria). Furthermore, the loss of oil revenues to corruption has been significant. The reality for Nigeria across decades of oil and gas production has been the inability of the political elite to create and then get behind a coherent strategy for economic, industrial and human development driven by oil revenues. This cause has been further hindered by Nigeria's failure to harness her natural resources locally and maximise Nigerian-skilled human capital in a productive manner.

It has been suggested that the oil and gas industry in Nigeria is an 'enclave sector' meaning that it is significantly disconnected from the rest of the Nigerian economy with very few multipliers beyond the rent that accrues to government and a small number of 'hydrocarbon stakeholders'. The industry has also failed to generate any significant levels of employment within the downstream sector. The focus has been, and remains on, the process of extraction but not on the wider downstream businesses that are a part of any thriving hydrocarbon-producing nation.

Nigeria warrants analysis for a number of reasons, not least because it is the largest nation in Africa (both in terms of population and economy) and its oil and gas sector is critical to the economic health of the country and its people. Since oil production began in the 1950s, successive governments have recognised the need to address skills and workforce development and have been pioneers in the emergence of a coherent concept around the participation of Nigerians in the business and the wider supply chain. However, the story here is not one of achievement but rather of an ongoing struggle to create – and to implement – the mechanisms by which Nigerians can play an active and meaningful part in an industry that has been and remains the cornerstone of their economy.

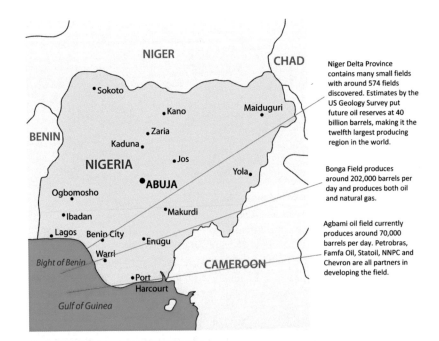

Niger Delta Province contains many small fields with around 574 fields discovered. Estimates by the US Geology Survey put future oil reserves at 40 billion barrels, making it the twelfth largest producing region in the world.

Bonga Field produces around 202,000 barrels per day and produces both oil and natural gas.

Agbami oil field currently produces around 70,000 barrels per day. Petrobras, Famfa Oil, Statoil, NNPC and Chevron are all partners in developing the field.

COUNTRY FACTS

- Nigeria has a population of over 170 million, making it the most populous nation on the African continent.
- An economic study conducted in 2009 by The Economist revealed that the oil sector accounted for 70–80% of federal government revenue (depending on the oil price), around 90% of export earnings and around 25% of GDP.
- According to the US Energy Information Administration, Nigeria was the fourth largest supplier of crude oil to the United States in 2010[1] (although due to changing dynamics in the US market, Nigeria stopped exporting oil to the United States in 2014).
  As of 2008, the extractive industries employed just over 93,000 people in Nigeria (representing 0.15% of the total working population).[2]
- A recent study suggested that Nigeria is one of the least efficient producers of oil on the planet[3]; in a low oil price environment, the challenges this creates for the Nigerian state and economy are immense.
- Nigerian-proven oil reserves were estimated to be 37 billion barrels in 2011.[4]
- Nigeria has an estimated 159 trillion cubic feet (Tcf) of proven natural gas reserves, giving the country one of the top 10 natural gas endowments in the world. Due to a lack of utilization infrastructure, Nigeria still flares about 40% of the natural gas it produces and reinjects 12% to enhance oil recovery.[5]
- The language of instruction in Nigerian institutions is English.
- The Ministry of Education is the government body charged with the duty of regulating procedures and maintaining education standards across the education sector.[6]
- Nigeria has 26 federal universities, 17 federal polytechnics, 25 state government universities and 26 state government polytechnics. There are also 23 private universities and a number of private polytechnics.[7] An additional 9 new private university licenses were awarded in February 2015.
- Access to higher education is severely limited and demand far outstrips the availability of places (about 1.5 in 10 applicants is successful for universities, with the figure rising to 1 in 4 for polytechnics).[8]

---

1. US EIA Statistics.
2. Jesse Salah Ovadia, *Indigenization versus Domiciliation: A Historical Approach to National Content in Nigeria's Oil and Gas Industry.*
3. Deutsche Bank study, 2014.
4. US Energy Information Administration, August 2011.
5. *Development of Nigeria's Oil Industry.* NNPC Website, http://www.nnpcgroup.com/.
6. United States Diplomatic Mission to Nigeria Website.
7. UKTI Nigeria.
8. Ibid.

THE HISTORY OF OIL AND GAS IN NIGERIA

Ship with oil rig in the background in the Nigerian capitol Lagos. (© *Shutterstock, vanHurck.*)

Nigeria's oil story began in 1908, when a German entity, the Nigerian Bitumen Corporation, commenced exploration activities in the Araromi area, West of Nigeria. The outbreak of World War I in 1914 put a pause on this activity with oil prospecting resuming in 1937, when Shell D'Arcy (the forerunner of Shell Petroleum Development Company of Nigeria) was awarded the sole concessionary rights covering the whole territory of Nigeria. Their activities were also interrupted by World War II, but resumed in 1947. The first commercial discovery of oil was in 1956 at Oloibiri in the Niger Delta 65 miles west of Port Harcourt.[9]

In 1958, the first year of production, average production was 5100 barrels a day.[10] During the early years of oil production in the Niger Delta, Nigeria was still a British colony with a largely agrarian economy. As oil revenues began to flow, the country found itself as a newly independent nation with money to invest. Despite the negative impact of the Nigerian Civil War between 1967 and 1970, oil revenues were the driver for significant investments in infrastructure (including schools, roads and hospitals). Revenues were also used to build the country's first oil refinery in Port Harcourt, a project that was commissioned in 1965.

Nigeria's economic development has, for many years, been inextricably linked (and overly reliant on) the rise and fall of the international oil market. Nigeria's vulnerability to oil price shocks is rooted in an overdependence on the export of crude oil. In 1961, crude oil accounted for just over 7% of total exports but by 1970, it had become the primary source of foreign exchange,

---

9. *Development of Nigeria's Oil Industry.* NNPC Website, http://www.nnpcgroup.com/.
10. Jesse Salah Ovadia, *Indigenization versus Domiciliation: A Historical Approach to National Content in Nigeria's Oil and Gas Industry.*

accounting for nearly 64%.[11] By 1979, following the Arab oil embargo against the USA, petroleum sales had completely overshadowed non-oil exports (cocoa, timber, cotton, rubber, etc.), accounting for around 95% of the country's export earnings.[12] During the peak of the 1970s oil boom, Nigeria's premium crude, known as 'Bonny Light', fetched $40 a barrel. During the oil glut of 1982 the official price of Bonny Light tumbled to below $10 a barrel. This had a catastrophic effect on the Nigerian economy and set the pattern for future economic crises as oil prices fluctuated.

During the Nigerian Civil War in the late 1960s, the ruling government took steps to ensure that control of oil would reside at the geographical and political centre of the country. Part of the Petroleum Act of 1969 mandated that ownership and control of petroleum should rest with the state and the federal oil minister had sole rights to grant oil mining leases. A system of revenue allocation was also established that would slowly reduce the benefit to the producing states from petroleum resources extracted in their territory. Between 1966 and the mid-1990s, the share of revenues that would flow back to the regions reduced from 50% to 3% (although this was later raised back to 13% in 1999).[13] This centralised model created economic imbalance, reduced the ability of states to invest locally and funnelled oil revenues to central government where corruption became rife.

By the early 1970s, petroleum revenues were flowing into the federal government with some commentators noting that the state was, at this time, struggling to find ways to spend the oil windfall. Oil revenues during the 1970s ushered in an era of massive government spending with some focus on infrastructure, schools and other education institutions.[14] In 1971, the country joined OPEC.

In the decades that have followed, Nigerian oil extraction has suffered from bureaucracy, inefficiency and incompetency driven largely by a lack of long-term thinking amongst the political leadership. The ongoing effects of militancy in the Niger Delta have also damaged the industry with the theft of crude oil becoming widespread. The failure of the government and the industry to build mechanisms by which local communities in oil-producing areas benefit from the activities of the energy sector has often been cited as a principal cause of militancy and widespread criminal practices in the region. In spite of these ongoing issues, official Nigerian oil production remains high and the size and value of the industry dwarfs other oil-producing nations in the region.

The Nigerian National Petroleum Corporation (NNPC) was created in 1977 to oversee the regulation of the oil and natural gas industry. In 1988, the NNPC was split into 12 subsidiary companies to regulate the subsectors within the industry.[15] The Department of Petroleum Resources, within the Ministry of

11. The Petroleum Sector by S.W. Petters; Online Nigeria.
12. Ibid.
13. Obi and Rustad (2011): 7.
14. Falola (1999): 143.
15. Nigerian National Petroleum Corporation Group Website; About NNPC.

Petroleum Resources, also has a regulatory function focussing on general compliance, leases and permits and environmental standards.[16]

A majority of Nigeria's major oil and natural gas projects are funded through joint ventures (JVs) between international oil companies (IOCs) and the NNPC, with NNPC as the majority shareholder. Other projects are managed through production sharing contracts (PSCs) with IOCs. PSCs are more common for deepwater projects as they typically involve more attractive fiscal terms than those in JV arrangements with a mechanism to incentivise the development of deepwater projects (which carry greater risk and require a larger capital investment). NNPC has JV arrangements and/or PSCs with Shell, ExxonMobil, Chevron, Total and Eni. Other companies active in Nigeria's oil and natural gas industry (and involved in PSCs) are Addax Petroleum and Statoil, alongside several smaller Nigerian companies.[17]

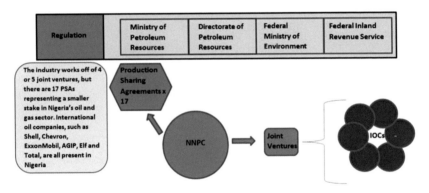

The Oil and Gas Sector in Nigeria.

As of 2015, the industry faces some critical challenges:

- Local refining of crude oil is at an abysmally low level due to the absence of functional refineries and sabotage on the ageing hydrocarbon infrastructure and pipelines. There is currently no commitment to turnaround maintenance or the overhauling of the facilities. Nor is there a plan to build new facilities.
- Poor management of Nigeria's Sovereign Wealth Fund has reduced the value of crude oil, which in turn has had an adverse impact on Nigeria's economy and budget. The Naira has been devalued to address the imbalance of the budget. Retrenchment is ongoing in the sector because of the anticipated revenue loss and other stringent financial measures that would result from reduced revenue.

---

16. The Petroleum Regulatory Agency of Nigeria Website; About Us.
17. US Energy Information Administration. *Nigeria Country Overview.*

- The Niger Delta today suffers from environmental pollution and degradation of land and waters that were previously sources of employment and revenue to the locals.
- Militancy has resulted from high unemployment and the neglect and nonengagement of young people. This has led to a significant reduction in the oil produced in Nigeria and has discouraged many IOCs from operating inland.
- There is significant uncertainty in the industry with many operators putting new projects and investments on hold. IOCs are divesting from onshore assets and focussing on investments in deepwater plays.
- Political stasis and the unwillingness of the National Assembly to pass Petroleum Industry legislation for the restructuring and regulation of sector is creating further uncertainty for operators and holding up foreign direct investment.
- More broadly, government policy relating to the oil and gas sector is considered by many to be unsustainable, unreliable and lacking in effectiveness.
- Many commentators also cite the ongoing problems with a lack of transparency in government regulatory agencies' dealings with industry operators and contractors as a key challenge that must be addressed if the industry is to move forward and reach its potential.

A BRIEF HISTORY OF EDUCATION, TRAINING AND WORKFORCE DEVELOPMENT

A group of local workers meet in Bonny Island, Nigeria 2006. (© *Shutterstock, Lorimer Images.*)

Out of a population of over 170 million people, 30 million are of school age. Of this 30 million, an estimated 10 million are not enrolled in school in Nigeria.[18] The Federal Government of Nigeria frames the objective of the

18. Figures from the US Embassy in Nigeria Factsheet.

Nigerian education system as a mechanism for effecting national development. The national philosophy towards education is based on the idea of developing individuals into effective citizens and giving equal educational opportunities to all citizens at primary, secondary and tertiary levels. Although the intention here is laudable, the reality for learners of every age and level in Nigeria falls far short of the government's lofty ambitions.

Educational institutions across the spectrum have suffered from the systemic and ongoing neglect of successive governments. Over recent decades, tertiary institutions have been plagued by riots and strikes resulting in – and reflective of – a decline in quality of the education system. Nigeria's education sector is not on a par with those of other nations (albeit that it shares many of the failings exhibited by other developing economies in relation to education and training) and there has been a failure to effectively implement international standards across the system or to develop a coherent, long-term approach to education and skills development.[19] A theoretical, didactic approach prevails in the higher education system and students experience a paucity of opportunities to apply theoretical knowledge in practical situations.

There is very little evidence of tangible results from the commitment made to achieving the Millennium Development Goals around education. Although education has been a priority issue for successive governments – and the rhetoric around education and skills development has been clear – the reality is that implementation of government plans has been patchy at best and the country has been left with a system that fails to support progression between education and employment. Graduates are largely viewed as unemployable by the industrial sector (albeit that this is a common refrain from employers around the world). Universities, polytechnics and technical and vocational colleges offer a wide choice of study areas but courses are predominantly academic in nature (even at technical/vocational colleges) and programmes of study fail to meet or take account of current industry standards, best practices or industry demands.

The poor quality and knowledge of academic staff is cited as a key challenge facing the higher education sector. Furthermore, student class numbers often run into the hundreds meaning that lecturers are under enormous pressure and are unable to do anything more than deliver their lectures. With demand for places far outstripping supply there is a scarcity of tertiary institutions offering relevant courses. For a nation with vast and diverse natural resources, vocational and technical colleges should be the first choice for many students. Unfortunately this is not the case as there are too few good and affordable educational institutions and a preference persists for students to go to university. Some commentators also identify a culture of negative work ethic as an issue

---

19. Teboho Moja, *Nigeria Education Sector Analysis: An Analytical Synthesis of Performance and Main issues* (World Bank, 2000).

facing the country with poor discipline rife and insufficient development of the behaviours and attitudes that are expected in the workplace.

A study undertaken by Volker Treichel, Lead Economist at the World Bank, set out the challenges that faced Nigeria during the later part of the 2000s in relation to skills and workforce development. These challenges included the following:

- Those in the formal education system favouring general (academic) education above vocational education
- Government strategies for skills development that are either a response to a crisis or temporary social measures to tackle unemployment
- Existing skills development programmes that lack appropriate funding, use outdated curricula and are short of qualified teaching staff
- Limited horizontal coordination across different ministries/agencies and vertically between federal and state authorities
- Implementation strategies not based on timely or accurate labour market information
- The absence of transparent quality assurance mechanisms
- The absence of mechanisms for recognition of skills acquired in the informal sector

In order to address these serious and systemic challenges, the study suggested the following:

- A re-prioritisation of the government's resource allocation to the technical and vocational stream of the education system
- A more strategic approach to the design of policies for skills development that are geared towards growth and anticipated growth areas
- The creation of a coherent national skills development strategy and better coordination of the implementation of activities in this area
- The promotion of public–private partnerships that are private sector led with the government providing a framework around standards and accreditation
- The establishment of more appropriate regulatory and institutional framework
- Speeding up the creation of a National Vocational Qualifications Framework

The impact of a failure to create coherent and effective skills development policies has been acute. Youth unemployment rates, particularly in rural areas like the Niger Delta, have risen steadily and created significant social problems and, with that, a rise in criminal activities and sectarian violence. Without the skills to support access to employment opportunities, poverty has become the norm for many school leavers. Within resource-rich areas of the country, the sense of powerlessness has been exacerbated by the comparative wealth of those working within the oil and gas industry, an industry that often seems uninterested in the plight of citizens within the communities where they operate. The subsequent issues of oil theft that have become a major problem in the Niger Delta can be traced, in part, to the disenfranchised populations in these areas and that, in turn, reflects the historic neglect of education and training at every level.

THE STORY OF WORKFORCE AND SKILLS DEVELOPMENT FOR THE OIL AND GAS INDUSTRY[20]

Overview

The story of how Nigeria has developed citizens to work in the oil and gas sector is one of limited success, and is a story set within a wider context of poor educational outcomes and a tertiary education and training system that labours under the weight of numbers and underinvestment. In considering how the government and the industry have approached this issue we need to look at the wider political context for the sector and consider efforts made to promote local participation at policy level. What we see here is a narrative that clearly demonstrates a vision for the oil and gas sector and an awareness of the wealth and job creation that it can stimulate. However, we also see a failure to back up policy on local content with coherent workforce and skills development actions. Local content law can set the target for achieving local participation but without talent and skills – and without the mechanisms to develop talent and skills over a sustained period – local content targets will remain notional goals or, even worse, will create sectors of the economy where those who are employed are not adequately skilled to operate safely and productively. Here we look at how the Nigerian government approached the issue of local content, what this meant for education and training and what other workforce development actions the government has undertaken to build local capacity and participation in the industry. We also look at the role that international operators have played and at the factors that are now shaping this dynamic.

The Struggle for Local Content

Nigeria has a long history of failed projects and failed policies in relation to effective economic and social development within and beyond the oil and gas industry. These policies can be broadly split into two categories:

- Policies aimed at supporting and promoting national ownership of the industry (in the belief that this will create a 'trickle down' of wealth to the wider population – these types of policies have often been referred to as 'Nigerianisation' or 'indigenisation' approaches)
- Policies aimed at increasing economic activity and job creation through knock-on effects of the oil industry – essentially, lengthening the supply chain, downstream business and so on

---

20. Ovadia J. S., *The Petro-Developmental State in Africa: Making Oil Work in Angola, Nigeria and the Gulf of Guinea* (London: Hurst, forthcoming 2015).

The concept of Nigerianisation was first given voice in policy through the Petroleum Act of 1969. This act, and those that followed over the subsequent three decades, was predicated on three principles:

- That policies would, in theory, generate revenues for the Nigerian government
- That policies would empower Nigerians by generating new opportunities for employment and wealth creation within the sector
- That policies would lead to Nigerians being in positions of power and strategic influence within key industries

The principles were built around the objective of ensuring Nigerians assumed management roles across the industry and that the day-to-day running of the industry was the responsibility of Nigerian nationals. This meant that ownership was the key and Nigerians needed to be educated, trained and developed in order to achieve the objectives of the Nigerianisation agenda. The Petroleum Act of 1969 addressed these aspirations by first setting rules around the ownership and management of licenses to ensure the participation of Nigerian citizens and companies. The Act also set ambitious targets for the transitioning of senior positions within operating companies to Nigerian citizens over a set period of time. This part of the legislation defined the requirement of operating companies to clearly set out plans for the recruitment and training of Nigerians as part of their in-country operations and to submit twice-yearly reports on their progress against these plans. More specifically, the 1969 Act stipulated that, within 10 years of the granting of a license, the operating company would

- employ at least 75% Nigerians in managerial, professional and supervisory roles
- employ upwards of 60% of Nigerian citizens within each of those grades
- reach 100% employment of Nigerians in skilled, semiskilled and nonskilled positions

The effectiveness of this policy was, in large part, related to the procurement and employment activities of the National Nigerian Petroleum Company who exerted considerable control across the industry. Despite this – and despite a clearly articulated plan for local content and local participation – the ambitious goals of the 1969 Act were never achieved.

One of the main reasons for failure in regard to the issue of business ownership related to the practice of 'fronting' whereby an expatriate business owner would sell their business to a Nigerian but retain effective control. This meant that, although Nigerian businesses were notionally engaged in oil and gas exploration and production and were able to access opportunities within the industry and the supply chain, the real power lay with foreign owners and investors who were able to control activities and take revenues out of the country. Further challenges to the policy of 'Nigerianisation' were recognised as being the result of a lack of skills, skills development programmes and inadequately qualified workers able to fulfil skilled and managerial positions across the industry (a struggle that persists to this day).

During the mid- to late-1970s, the Nigerian government recognised the continued failure of their policies in relation to ownership and workforce nationalisation and took steps to increase state ownership of the IOCs operating in Nigeria at that time. By 1979, the government enjoyed around 60% equity participation in the IOCs active in Nigeria. Despite their share of equity, control of these companies remained outside of Nigerian hands. This was seen primarily as a consequence of the lack of Nigerian management talent available to fulfil senior positions and the inability of the state to create education and training programmes and employment pathways into these senior positions. With the government increasingly relying on oil revenues to fund their legislative programmes – and with corruption and deal-making an ongoing backdrop – the desire of the political class to promote Nigerian participation in the industry was tempered by a recognition that the sector needed to be run and operated by those with the right skills, experience and expertise. Having failed to invest in education and training, the country suffered a paucity of talent coming into the industry and therefore had no choice but to support the continued dominance of a largely expatriate workforce. The unwillingness of successive governments to actively enforce policies of local participation needs to be viewed within this light.

A number of other approaches have been subsequently adopted in an attempt to support Nigerian participation in the sector (although, arguably, none have addressed the ongoing challenge of local workforce and skills development). During the 1990s, the government began to sign Production Sharing Contracts (PSCs) with IOCs in order to ensure Nigerian involvement in exploration and production activities whilst maintaining the viability of what were often costly projects. Although these agreements were not always made public, it was understood that every PSC would contain clauses requiring the preparation and implementation of plans and programmes for the education and training of Nigerians for all job classifications in accordance with the Petroleum Act of 1969 (which was still part of national law). These agreements would also set the parameters for the involvement of local companies in the provision of services to the industry. Once again, the issue was one of enforcement. Commentators view these agreements as being well thought out and sensible in their aims but largely ineffective at achieving their goals. Too much of the responsibility fell to IOCs who were unwilling or unable to meet the terms of their agreements and were relaxed about the consequences to their business. This failure reflects significantly on the inability of the key participants across the industry to address skills development issues and to create a system of education and training that matched the ambitions of the government in regard of workforce nationalisation.

Further evidence of this issue can be found in the implementation of the discretionary allocation policy (DAP) during the 1990s. This policy was established to award oil prospecting licenses to Nigerian entrepreneurs (ahead of their international competition). By 1998, 38 licenses had been awarded to indigenous companies but the scheme was quickly exposed as being largely an exercise in political patronage with those receiving licenses typically the friends and associates of senior politicians. In assessing the success of the DAP, the then director of the Department for Petroleum Resources cited the inability of the companies awarded contracts to 'muster the required know-how to deliver' as the key reason for the failure of the policy. Once again, an innovative idea for promoting workforce nationalisation floundered on the twin rocks of a lack of local talent and political corruption. The consequences of ineffective or unenforceable policies increasingly impacted on the communities that shared space with oil and gas producers. A lack of local employment was fuelling restlessness and disillusionment amongst young people, local communities were increasingly frustrated at the lack of development and investment, oil- and gas-producing areas were witness to capital flight and there were seemingly no linkages between the oil and gas industry and other industrial sectors. Something had to change.

By 2000, the historic failures of successive policies had coalesced into the growth of a recognisable movement for change in relation to the promotion of local participation in Nigeria's oil and gas sector. The continued reliance on the importing of goods and services to support economic activity coupled with ongoing systemic failures in the strategy and implementation of education and training policy had resulted in an unbalanced economy that struggled to create jobs and an oil industry that was creaking under the weight of its own economic responsibility for the country as a whole. Following the establishment of a National Committee on Local Content Development – supported by the Obasanjo administration – the Committee undertook a comprehensive study of the state of the industry and concluded, in early 2002, that Nigerian participation in the upstream oil and gas industry stood at less than 5% meaning that 95% of the estimated US$8 billion spent every year across the industry flowed out of the country. This statistic is even more shocking in light of the continuous efforts of successive governments to address local content (beginning with the Petroleum Act of 1969, more than 30 years earlier). The establishment of the National Committee on Local Content Development was the precursor to the writing of new legislation that would govern future local content policy in Nigeria. This policy was heavily influenced by a report produced by the Norwegian development agency Norad, which, amongst other things, recommended the establishment of a legal framework around local content policy. However, further political issues hampered the creation and implementation of this framework with changes in personnel within the ruling party leading to delays and resulting in the bill only being passed into law on the April 22, 2010.

In 2010, the government (after years of ineffective policy and political lethargy) instigated the Nigerian Oil and Gas Industry Content Development Act (NOGICD). When the new local content bill arrived, it outlined in the schedule of the Act targets for the level of Nigerian participation in the oil and gas sector with a headline target of 70% local participation being the most eye-catching. The Act also stipulated a maximum of 5% expatriate participation in senior management positions and minimum targets for specific areas of business including engineering, fabrication, materials and procurement, services, research and development, shipping and logistics (amongst many others). The Act also looked in detail at how Nigeria could transition towards a country that had much greater ownership of all aspects of the oil and gas supply chain and set out the key areas for attention. The field of front-end engineering and design was identified as a critical aspect of the local content challenge. This was an area of the sector that involved a large out-of-country spend, was key to capacity building and technology transfer and was a stepping stone to other parts of the wider supply chain. In this area – as with other areas of the sector – the committee charged with developing the Act set out current capacity against required capacity across the industry. This created a degree of awareness around the specific skills needed and formed the basis of a more coherent workforce development plan. The committee was also keen to generate a full audit of current Nigerian human capacity – something that would have been of significant value to workforce planning efforts – but this never materialised.

A critical aspect of the Act is the direct distinction that is made between workforce nationalisation (so-called 'Nigerianisation') and 'domiciliation' (or 'doing it in Nigeria'). The policy – and those who advocate it – specifically talks about focussing on the latter. The idea is not to drive international companies to employ more locals. Rather it is to ensure that every aspect of every project is fulfilled by a local company wherever possible. This ensures both the ownership that has been an aim of policy for 45 years and, as a by-product, is likely to ensure the nationalisation of the workforce (with a majority of Nigerian companies largely employing Nigerian nationals). On signing the legislation through, the government claimed that the impact could extend to 30,000 new jobs for Nigerian nationals alongside a positive impact on GDP growth. However, the critical challenge of skills and workforce development remained the elephant in the room. How could Nigerian companies fulfil contracts in the highly technical oil and gas sector if they did not possess the skills, the talent, the experience and expertise to operate safely and productively?

The progress of the 2010 Nigerian Oil and Gas Industry Content Development Act was stalled, to an extent, by opposition from established IOCs (although the exact details of this opposition are hard to verify). What is clear is that the Act was coherent in its approach and there was (and remains) the political will to implement and enforce the policy on the ground. This would inevitably mean changes for those international companies operating in Nigeria and whilst, on the surface, many companies seemed positive about the proposed changes in legislation, the ability of IOCs to operate in Nigeria with relative impunity was under threat and this was clearly recognised by those with commercial interests in the country. Anecdotally, the attitude of many of these companies at the time was 'how can we do as little as possible' in relation to Nigerian participation. As the law was passed in 2010, IOCs were vocal in their criticisms of the level of reporting that was required to comply with the new legislation. That said, the response of some IOCs to the implementation of the Act was broadly positive – reflecting that of many business groups and interested parties across Nigeria. There was a collective recognition that the plan would be of real benefit to Nigerian citizens and to the communities where oil and gas activities were prominent but that these benefits would only accrue if everyone took responsibility for implementing the policy. The slow but steady progress of IOCs towards a greater degree of corporate responsibility coupled with a growing understanding of the clear economic benefits of employing locals over expatriates further fuelled cautious optimism amongst international companies operating in Nigeria.

Under the NOGICD Act, operators in the industry are obliged to implement succession plans for Nigerian understudies. However, the processes and procedure for effective knowledge transfer in the form of global exposure, mentoring, coaching, continuous professional development and performance management are yet to be fully implemented.

Ultimately the success of the 2010 Nigerian Oil and Gas Industry Content Development Act will be measured by the degree to which Nigerian companies and Nigerian citizens are actively engaged in economic and industrial activities related to the oil and gas industry. Figures supplied by the Nigerian Content Development and Monitoring Board in 2014 state that engineering in the oil and gas industry is now completed 90% in country and that fabrication of field development facilities now has 50% of the tonnage completed in Nigeria.[21] This represents some measure of success for the new policy. Nigerian authorities have made clear statements about the focus that they will bring to the task of evaluating the quality and impact of Nigerian companies. Simply ensuring Nigerian participation is not an end in of itself. The productivity and economic success of Nigerian companies winning contracts across the sector will define whether the policy is really working and this relies on Nigerian companies having access to suitably skilled technicians, operators, managers and leaders. Without investing in long-term skills development across the country, any policy of this nature is destined to fail, as previous policies have proven.

Further Government Efforts to Directly Address Skills Development for the Oil and Gas Industry

Beyond the pursuit of effective policies and laws governing local content and local participation, successive Nigerian governments have grappled with issues of workforce development for the oil and gas sector with mixed results. Here

---

21. *Review of the Nigeria Oil & Gas Industry* (Nigeria: PWC, 2014).

we look at some of the developments of note as well as reflecting general perspectives on the dynamic between government and industry in Nigeria.

The Establishment of the Petroleum Training Institute, Effurun

Within a context of growing awareness of the importance of promoting local participation across the oil and gas sector, the Nigerian government set about addressing the skills development challenges it faced by implementing a number of initiatives. First amongst these was the establishment of the Petroleum Training Institute (PTI), Effurun in 1972. The Institute was established under a bilateral agreement between the Federal Government of Nigeria and the Soviet Union. As part of the agreement a number of Soviet experts were drafted in to support the delivery of programmes and training equipment was provided during the early stages of the Institute's operation. The number of Soviet lecturers and instructors gradually decreased over time as local capacity grew.

The Institute was established to fulfil a number of related tasks and responsibilities. These were defined as follows:

1. To provide courses of instruction, training and research in oil technology and to produce technicians and such skilled personnel normally required for oil production
2. To arrange conferences, seminars and study groups relative to the offered fields of learning
3. To perform such other functions as in the opinion of the governing council may serve to promote the objective of the Institute, including the making of regulations for entry into any of the courses approved by the Institute, the duration of such courses and their academic standards and the recognized equivalents of such certificates and diplomas that the Institute may award.[22]

The PTI remains today a key part of the Nigerian government's efforts towards building a national skills base and it has received considerable funding and support from the Petroleum Technology Development Fund (PTDF) – see below. Its focus is on the delivery of industry-relevant programmes awarding higher national diplomas and certificates in key skills areas like petroleum engineering, mechanical engineering, electrical engineering, petroleum business studies and, at certificate level, welding and commercial diving. PTDF has provided strategic guidance and funding in order to support the Institute in meeting the rapidly changing skills demands of the oil and gas sector (with an increasing focus on highly technical areas like deepwater oil exploration and production creating new challenges in terms of education and training). PTDF has invested heavily in the Institute through the provision of modern technology

---

22. Momoh, Hadiza Tijjani Sule (Mrs) *Nigerian Local Content Act: The Role of the Petroleum Training Institute, Effurun in Human Capital Development.*

and infrastructure and this commitment demonstrates the important role that the PTI plays in making real the strategies and ambitions of PTDF.

However, the story has not been one of unmitigated success. Over time, the PTI has come in for criticism on a number of fronts. There have been suggestions that the quality of teacher and trainer knowledge often falls behind that required by a fast-moving industry and greater emphasis needs to be given to ongoing staff training. There is also evidence that the Institute has not managed to forge strong enough relationships with Nigerian and international companies operating in the oil and gas sector. As a consequence, access to hands-on training opportunities is at a premium and, without this vital component, the programmes offered at the Institute are diminished in the eyes of potential employers. The propensity of IOCs to develop their own in-house training initiatives (or to simply transport those they want to train to overseas locations) would suggest that the quality of PTI graduates – at least in the eyes of some employers across the oil and gas sector – is, unfortunately, questionable.

The Establishment of the Petroleum Technology Development Fund or PTDF

One of the major constraints facing Nigeria's oil and gas industry from first oil was the lack of skills and technical know-how and the subsequent impact that this had on the ability of citizens to play an active part in this burgeoning industry. The shortage of required skills, attitudes, knowledge and competencies among Nigerian young people, particularly in southern part of Nigeria, led to high unemployment rates and, in turn, fuelled increases in poverty. The cost of gaining access to Nigeria's education system was out of reach for many in Nigerian society and this created high rates of illiteracy among Nigerian young people. Without the basic skills of reading and writing, the talent pipeline for all sectors – including oil and gas – was running dry. Furthermore, there were very few established education and training programmes that would effectively prepare candidates for jobs in the oil and gas sector. It is against this backdrop that the PTDF was given a mandate to create pathways to education, training and, ultimately, employment into the oil and gas sector.

The establishment of PTDF – and its continued support and evolution – represents one of the principle attempts by which successive Nigerian governments have sought to address workforce development needs within the oil and gas sector. The partnerships that exist between PTDF and the government agencies responsible for developing and promoting local content legislation demonstrate the awareness that has existed for many years of the critical connection between industrial and economic policy on the one hand and active engagement with skills development on the other.

PTDF was established by Act No. 25 of 1973 to replace the former Gulf Oil Company Training Fund Act for the purposes of training Nigerians for the Petroleum Industry. Between 1973 and 2000, PTDF functioned as a desk in the Department of Petroleum Resources (DPR), until September 2000 when it was made an independent and fully functional Parastatal under the administration of President Olusegun Obasanjo.[23] The stated mission of PTDF is to train Nigerians to qualify as graduates, professionals, technicians and craftsmen in the field of engineering, geology, science and management in the oil and gas industry and its vision is to serve not only as instrument for the development of indigenous manpower and technology transfer/acquisition in the petroleum industry but also to make Nigeria a human resource centre for the West African subregion in the oil and gas sector.

In a 2009 report by a former executive director of PTDF, the author refers to a 'Local Content Awakening' that occurred at the end of the 1990s and sets out the impact this had on PTDF (which had been operational since 1973 but, it is inferred, had failed to have any significant impact on skills development for the oil and gas sector). The report identifies that 'the Nigerian content development initiative requires an exhaustive systemic approach that would assess local content levels, identify constraints, develop clear policies and processes to stimulate growth as well as define clearly the roles and responsibilities of stakeholders'. Within this context, the newly revitalised PTDF was described as the 'capacity building organ of government'. The task facing PTDF was to first undertake an assessment of local content levels, and, in the process, identify the constraints that existed in relation to the availability of required competencies and capacities. In essence, PTDF needed to understand what the workforce requirements of the industry were and what current skills existed within the market. The second task was to develop a set of capacity building policies and programmes that would stimulate local content growth and support the renewed focus of the government around the participation of Nigerians in every aspect of the oil and gas sector.

At this time – prior to the 2010 Nigerian Content Bill – there was no comprehensive legislation guiding Nigerian Content although the federal government had set several targets (of 45% by 2006 and 70% by 2010) for Nigerian participation in the oil and gas industry and had issued 23 short-term directives to guide the implementation of government policy. One aspect of these directives was to 'set the compass' for the role that PTDF would play in terms of capacity/capability development programmes and how these activities would support wider efforts at promoting Nigerian participation in the sector. The key objective for the organisation was to stimulate and support job creation by developing a critical mass of highly skilled indigenous candidates who were suitably qualified to international standards and who were able to assume roles across the sector. There was also a wider remit around stimulating cross-sector linkages leading to the active participation of Nigerians in all sectors of Nigeria's economy.

---

23. From the PTDF Website: http://www.ptdf.gov.ng/index.php/aboutptdf/history.

In 2004, PTDF commissioned an industry-wide skills gap audit and survey – undertaken by Nigerian consultancy Lonadek – to identify inadequacies in skills provision across the oil and gas industry. The aim of this was to provide inputs for a framework that would guide the development of an effective education and training system that was able to redress inadequacies and that was aligned with the skills demands of the industry. The audit revealed that achieving the 70% Nigerian Content target by 2010 set out by government policy would involve action on a number of fronts including:

- better support for technology transfer
- increased focus on skills training
- establishing mentoring and apprenticeship schemes for design engineering works, fabrication, construction, service and maintenance
- upgrading of existing organizational and educational facilities

The report was the precursor to a concerted period of activity on the part of PTDF that was designed to create genuine impact and that would, it was hoped, reverse the decades of relative inactivity that had preceded the changes brought about in 2000. Efforts were focussed on the following:

- **Collaboration between PTDF and the Nigerian National Petroleum Corporation**
  To enable PTDF to deliver on its mandate in relation to the Government's national content policy, the organisation signed a memorandum of understanding with the NNPC; this agreement was labelled "Job creation through partnering initiative" and was the mechanism under which several programmes were developed and jointly executed. The principal objective of the initiative was to develop the skilled workforce in order to achieve the Government's Nigerian Content target of 70% by 2010. Underpinning this was a focus on task-oriented certification and accreditation programmes and the establishment of local apprenticeship and mentoring schemes.
- **Establishing the Engineering Design Training Program (EDTP)**
  The EDTP was conceived as a way of addressing skills shortages within the key area of engineering design (something that had been specifically highlighted in the PTDF skills audit). The EDTP gave focus to the training of candidates in a number of engineering design software packages as a means of creating the required skills bases to fulfil all current and planned design projects across the Nigerian oil and gas sector. Initial targets for the programme were to train upwards of a 1000 engineers every year. The EDTP programme was further bolstered by a 'post training attachment' initiative aimed at ensuring that graduates from the EDTP had clear pathways into employment and that their skills were fully developed and relevant to the job.

- **Enhancing the fabrication capabilities within the Nigerian oil and gas industry**

  In its skills audit, the PTDF identified the lack of fabrication capabilities as one of the single largest impediments to the development of capacities and the building of local content. To address this, they embarked on two related projects: enhancing the fabrication capabilities of engineering yards and utilising these enhanced capacities to deliver certified training for welders. The project to develop the capacities of fabrication yards in Nigeria was conceived as a partnership between PTDF, the NNPC and the Norwegian development agency Norad. The welders training and certification programme was designed as a follow-up to these efforts and set out to train and certify around 2500 Nigerians in various aspects of welding (with this also being identified in the skills audit as a skills area of high demand).

- **Institutional Capacity Development**

  Alongside what might be viewed as more technical/vocational activities, PTDF also set out to support the evolution of higher education institutions engaged in fields of study and research related to the oil and gas sector. To this end, they undertook a series of institutional capacity building programmes involving support for Nigerian universities. The focus of these activities was to support the upgrade of selected departments in a number of key universities including the following: University of Port Harcourt, Gas Engineering Department, University of Maiduguri, Geology Department, University of Ibadan, Petroleum Engineering Department, University of Nigeria Nsukka, Geology Department, University of Benin, Chemical Engineering Department, Ahmadu Bello University, Chemical Engineering Department, Usman Dan Fodio University, Petroleum Chemistry Department and University of Jos, Geology and Mining Department. Support activities revolved around improvements to infrastructure, laboratory equipment, books, IT facilities, power generating sets and water borehole facilities. A second phase of the upgrade programme followed successful completion of the first phase with an additional eight federal universities becoming recipients of similar help.

What was clear from this detailed plan was that post-2000, PTDF – as the principal mechanism by which the government of Nigeria was aiming to promote and support skills and workforce development for the oil and gas sector – had been given renewed purpose, increased power and a wider remit. By framing their work with a comprehensive study of skills demands set against planned projects and current provision, the organisation was able to construct programmes and initiatives that were specifically geared around addressing the challenges the country faced. The success of these programmes and initiatives is less easy to measure however. Some progress was certainly made during this period. That said, if we consider a recent assessment of the status and impact of PTDF, we may question the distance that the organisation has actually travelled. At a meeting held in Houston in 2014, the then Executive Secretary of PTDF set out the key constraints that the organisation faced in providing Nigerian technicians and artisans with the quality training they need in order to be able to participate effectively in oil and gas projects and ultimately enhance Nigerian Content in the industry. These constraints were as follows:

- A lack of well-qualified instructors to train students and impact knowledge
- Inadequate modern training facilities for technical and vocational training
- Training curricula not in congruence with the requirements and needs of the oil and gas industry
- The preference of young people to pursue a university degree as against technical/vocational qualifications
- Weak systems that cannot sustain quality technical and vocational training.

Set against the backdrop of decades of effort – both in relation to the promotion of Nigerian participation in the industry and the investment in education and training – and put within the context of the post-2000 investment in PTDF as an organisation, these wide-ranging barriers represent a searing indictment of policy and practice in regard of meeting the skills and workforce development demands of the oil and gas industry. The Executive Secretary then went on to set out the required measures necessary to address the constraints outlined. These included:

- The development and implementation of a national strategic framework for technical and vocational training and education.
- Improvements in the teaching capacity and capability of vocational education instructors on a regular basis through capacity building in relevant areas.
- A thorough proficiency audit undertaken to ascertain the level of teaching and training competence current operating within the sector.
- The provision of requisite facilities and infrastructure in technical/vocational institutions to enable them to deliver on their mandate.
- A revision of the curricula of technical and vocational institutions to meet twenty-first century industry demands.
- Greater collaboration between government and international organisations as well as the private sector to develop technical/vocational training in-country.

It seems that as with the efforts of successive governments to implement a meaningful and achievable policy towards local participation in the oil and gas sector, parallel efforts to promote and support skills and workforce development have been limited at best in their impact. The industry remains largely staffed by nonnationals and the broader context for education and training in Nigeria remains fragmented, underfunded and lacking in a cohesive strategy.

Wider Perspectives on the Role of Government

The Nigerian government has historically played a role in encouraging nationals to compete for state and federal government scholarships within a meritocratic system that recognised and supported the best brains in Nigeria. This led to a good crop of graduating professionals, some of which became leaders within the industry. However, the Government approach has become more political, tribal and nepotistic and this has had a negative impact on developing human capital through the award of scholarships.

Investment in some higher education institutions has been influenced by the Government with focus given to projects that build and upgrade institutions. However this has been more about the construction of buildings and less about driving capacity, capability and competence development. The PTI, Effurun; the Federal Polytechnic of Oil and Gas, Ekowe and the Federal Polytechnic of Oil and Gas, Bonny Island; the National Centre for Skills Acquisition and Training in Oil and Gas in Greater Port Harcourt and the National College of Petroleum Studies (NCPS), Kaduna have all been recipients of this support.

In order to bridge the knowledge, competency and engagement gaps, the Nigerian government has introduced various measures – in addition to the enactment of the Nigerian Oil and Gas Industry Content Development (NOGICD) Act – to train and develop her citizens. These include the following:

- The Government has, through the Nigerian Immigration Service (NIS), regulated the influx of expatriates in Nigeria, through limiting the issuance of temporary work permits and other immigration permits required to work in Nigeria. The NIS currently implements a policy which requires employers to prove that the skills for which an expatriate is hired cannot be found in Nigeria and also mandates companies with expatriate quotas to submit monthly reports and a succession plan for Nigerian understudies, something that is also addressed in the NOGICD Act.
- The Government has introduced several schemes to assist Nigerians in the area of Training and Development. These include the Nigerian Content Human Capital Development Module, which is an integral part of any project executed in Nigeria; the PTDF Local Scholarship Scheme (LSS), which affords hundreds of Nigerian to embark on paid training programmes; the Industrial Training Fund (ITF) and Students Industrial Work Experience Scheme (SIWES) – these afford Nigerian undergraduates to undergo internships in companies operating in the sector in order to gain practical and industrial experience; the Education Trust Fund (ETF), which provides funds for the upgrade of tertiary institutions

A common criticism of government involvement in the sector is that Government-owned institutions and vocational centres do not have the required manpower in terms of experienced and certified lecturers in order to have the requisite impact on knowledge and skills transfer. The link between the relevant associations, societies and institutions that are critical to developing certified oil and gas professionals, vocationally trained individuals and technicians is lacking. The curricula that these institutions implement are out of date and not reviewed regularly in line with international standards. And the Nigerian University Commission has failed to forge partnerships with global bodies and local bodies for the accreditation of their programmes.

The Efforts of International Companies to Build Education and Training Capacity

As an industry historically dominated by international operators and service companies, their efforts merit attention in this regard. The IOCs and other service providers have played a significant role in funding scholarships (usually at secondary and undergraduate levels), offering internship opportunities, and in funding research and development activities.

Operators have also played some role in the improvement of local education capacity. Chevron, for instance, has been involved in upgrading the curriculum of tertiary institutions and has committed to volunteering their staff to lecture students (from a practical perspective). A pilot scheme was run with the University of Lagos (The Faculty of Engineering, University of Lagos, Human Capital Development Initiative) and Chevron intends to extend this to other Universities in various parts of Nigeria.

IOCs and service companies are also involved in a number of community development initiatives, which are focussed on educating and empowering host community citizens with relevant skills and knowledge required to operate in the industry. These activities include vendor/supplier development programmes, vocational skills acquisition training, entrepreneurship training and the award of grants to aid development.

The private sector has also contributed by implementing scholarship schemes, internship opportunities and community development programmes. Examples include the Niger Delta Development Commission technical and vocational educational training schemes, the Chevron Youth Training Scheme, the Shell Technician Training Scheme, the Amnesty Training Programmes for restive youths in the Niger Delta, The Energy Institute Young Professional Network (YPN), The Lonadek Vision 2020: Youth Empowerment and Restoration Initiative, Energy Institute, Lonadek and PennWell Youth Programme and the Afren School-to-Work Initiative. Other companies have also played a role in skills development including Total, Schlumberger and ExxonMobil.

Whilst many of these initiatives are to be applauded, they typically focus on graduate and postgraduate programmes, scholarships and education experiences that take learners overseas. They operate largely in isolation of the wider system of education and training and do not address the longer-term need for local capacity development. Furthermore, scholarship schemes are inherently limited in their reach with only a handful of the most talented students benefitting. A majority of these cases do not involve partnerships with local education and training providers which demonstrates the distance there is to travel in relation to local participation in this part of the oil and gas value chain.

When we assess the role of international companies there is, in fact, evidence that these companies have hindered progress on skills and workforce development. The reputation that international operators have in Nigeria is, in many cases, poor and this goes back to the early days of oil and gas production during which time Nigerian participation was acknowledged but never embraced. Although evidence of explicit workforce policies that favour expatriates over nationals is slim, it is clear that IOCs and other international operators failed to support the government in its efforts to build local capacity across the sector. These companies – who are part of JVs and PSAs with the NNPC – are accused of a litany of offences, corrupt practices and neglect including:

- A lack of investment in localised research and development
- No deliberate effort for succession planning and shadowing of expatriates
- A culture of poaching of staff amongst major companies (rather than developing or nurturing new talent)
- Indiscriminate employment portfolios
- A lack of compliance with existing national content laws
- Core productions and services being routed abroad and undertaken by foreign experts
- Selecting beneficiaries based on place of origin and not capability
- Discriminating between foreigners and Nigerians
- Inadequate support to foster international exposure for locals
- Low drive on technology transfer for growing local capacity
- Corporate espionage
- Corrupt practices
- Refusing to collaborate with PTDF and engage the talent who are trained and empowered with skills to meet the demands of the industry.

In analysing the contribution of international companies, it is difficult to judge the voracity of the arguments on both sides. On the one hand, the government would claim that IOCs and service companies act in isolation and without regard for local content requirements. On the other hand, the industry would point out that they lack a coherent framework within which they can contribute to workforce and capacity development. There is an element of truth in both positions and it will take the efforts of both to change the dynamics.

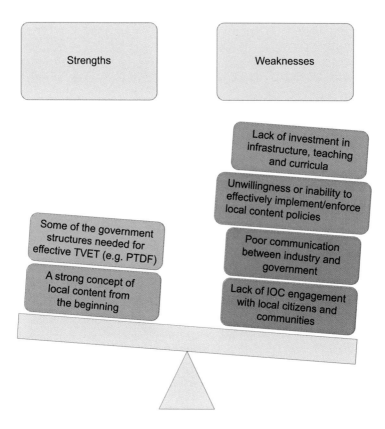

THE FUTURE

The sheer scale of Nigeria's oil and gas reserves means that the industry will continue to be an enormously important part of the economy for many years. The challenge for those governing the country's complex and somewhat troubled industry will be to increase Nigerian participation, connect the activities of the industry to the communities within which these activities take place and ensure that exploration and production activities are more efficient, safe and productive. This future will depend on a number of key factors:

- The government needs to reframe it's role within the industry and should become the regulator assuming all necessary powers to enforce better standards and ensuring adherence to local participation laws.
- Part of this will be to reduce rate disparity so that locals and expatriates who have the same skills and expertise can earn equally and focussing on performance-based pay.
- Across the board, a coherent concept of continuous professional development (CPD) needs to be promoted alongside the development of a culture where self-motivated personnel have a thirst for knowledge, skills and know-how.

- Reform of Nigeria's tertiary institutions should focus, in part, on the ranking departments in terms of employability. The regulation of these institutions by government has had the effect of politicising the citadel of knowledge and has reduced the academic capability of the institutions and diminished the learning experience.
- There is also a need to develop better mechanisms for the funding of University department chairs and funding grants to support research and development programmes that increase collaboration between the academic community and the private sector.
- To support this, enacting laws that create an enabling environment for research and development within tertiary education would create a stronger higher education sector that could support the industry in future.
- Companies need to be encouraged and supported to put succession plans in place alongside expatriate quotas to ensure that expatriates transfer knowledge and technology to local employees in the shortest possible time.
- Organisations that support the interface between industry, government and education should be supported and encouraged. The Oil and Gas Training Association of Nigeria – a government quango that is led by industry – is one such organisation.
- There needs to be much greater focus on developing local education and training capacity so that training can be undertaken in Nigeria. This will build a legacy of education and training infrastructure that will outlast the hydrocarbons.
- Better and equal collaboration between industry and government has to be a big part of the future – too often, the government has acted in isolation of the industry. It must listen, reach out and create forums for partnership and collaboration.
- The government also needs to become more focussed on enforcement of local content laws and on supporting local provision. History shows us that IOCs operating in Nigeria have often avoided investing in local education and training as they see this as a cost rather than a business benefit. The government needs to be proactive in bringing these companies to the table and ensuring they engage in a meaningful way.
- This effort should be part of a wider approach to developing and reinforcing collaboration between industry and government. If this dynamic can be more effective and sustainable, the impact on the industry could be transformative.

Ultimately, a multifaceted and multidimensional effort is required by all parties across the Nigerian oil and gas industry. The Nigerian Oil and Gas Industry Content Development Act stipulates that operators should give Nigerians first consideration for employment and training on any project being undertaken whilst ensuring that a reasonable number of personnel are employed from local areas. In the event that Nigerians are not employed because of their lack of training or skills, operators are required to make reasonable efforts to support the development of these skills locally and to offer the necessary training to do this.

The NOGICD Act represents a ray of hope for nationals. However, it will all depend on the degree to which the Act is enforced. There is a belief that greater workforce domestication can be achieved if the NOGICD Act is effectively implemented. Although efforts are being made to encourage compliance with the law, more needs to be done. Random inspections should be conducted, and stringent penalties should be imposed on companies that employ foreigners in contravention of the law.

If the new Act (unlike those that preceded it) is effectively enforced – and if the government and its industry partners can build and sustain a meaningful dialogue, then the future for Nigeria could be bright. The country faces a period of opportunity – by training and developing Nigerians in the right way, the country could begin to export talented, competent oil and gas employees to economies all over the world and thereby rewrite the story of Nigeria's oil and gas workforce.

The Getenergy View

- Nigeria offers us an apposite example of how it takes more than just policy to impact on workforce development. From the late 1960s, successive Nigerian governments were fully aware of the economic and employment benefits that could accrue from the emerging oil and gas sector. However, successive governments failed to enforce legislation meaning that targets for local participation remained just that.
- Alongside the lack of enforcement, we can also see that efforts at local participation will fail if the basic talent and skills pool is not there to support the policy. Regardless of the willingness or otherwise of international operators, they cannot be expected to employ local people and contract local companies if the skills and talent is not there. In an industry that relies on the twin pillars of safety and productivity, local participation has to be predicated on the availability of suitably skilled employees. Countries that want to achieve anything in regard to local content need to develop a coherent and connected education and skills policy to sit alongside economic and industrial policy.
- Local content rules can be a stimulus to oil companies investing in building local capacity. However, without a coherent approach to governing the whole sector – and without the right mechanisms in place to coordinate across the training and education landscape – these efforts are likely to be isolated and can end up having very little impact on the wider capacity of the education and training system.
- Strengthening the capacity of higher education institutions – and ensuring that university graduates are developed to support the industry – is a worthy and worthwhile cause. However, the real challenge lies around technical and vocational study with the majority of direct and indirect jobs created for technicians and artisans. It is tempting for governments and IOCs to focus on university-level education as this is often a 'quick win'. Getting to grips with technical and vocational training is much more challenging but no less vital.

Continued

The Getenergy View—cont'd

- Establishing dedicated oil and gas training colleges (like the PTI) should be applauded and these institutions can act as beacons of best practice to the education and training sector as a whole. That said, publicly run facilities need to establish strong and sustainable links to national and international employers to support practical experiential learning and to provide clear pathways into employment. Otherwise, these initiatives will provide access to education and training to local citizens but many may not find employment in what is a highly demanding, technically challenging industry.
- One of the biggest challenges around the success of education and training initiatives is sustainability. For programmes to be sustainable they must be based around a coherent partnership between industry and education and funding models need to be clearly defined. Without this in place, even successful projects can fail to gain longer-term traction.
- The industry moves fast. For too long in Nigeria, the IOCs led the dance with the government lagging behind. This is, in part, a result of short-term political thinking. Without a clear long-term vision – and the ability and will to drive this vision forward – IOCs will typically exploit opportunities for their own benefit. It is incumbent on the government to react positively to the changing dynamics of the industry.
- The oil and gas sector in Nigeria has too often become politicised with successive governments failing to address systemic issues, particularly in relation to local content. There is a need to approach the oil and gas sector as a business and for the government to recognise the critical role of the private sector in managing and driving forward exploration and production activities.
- It should be recognised that Nigeria has had some success in nationalising their industry. Although it is difficult to verify figures, there is a proportion of the oil and gas workforce that is Nigerian. However, the country also has one of the highest operating costs of any country in the world. This goes back to the need to ensure that employing local people is not an end in itself but is a desirable business decision that is based on competence, safety, productivity and cost.
- The fact that there are Nigerian expats around the world working in the oil and gas sector is testament to the progress that has been made within the industry in Nigeria. However, there is evidence that talented Nigerians (particularly those trained and educated by IOCs) often choose to leave Nigeria and work elsewhere. The drain of talent from the industry in Nigeria offers a further challenge to the progress of workforce nationalisation.
- We are at the start of a new era in relation to workforce and skills development for the oil and gas sector in Nigeria. Under the recent local content Act, local content rules are getting tougher and the government is pushing for more to be done. Although this should be applauded, the targets set have to be realistic and, crucially, the policy needs to be backed up by longer-term thinking and action around building a world-class, sustainable education and training infrastructure that can be the cornerstone of Nigeria's oil and gas future.

Case Study 3

Brazil: Global Success through Local Capacity Building

Chapter Outline

INTRODUCTION

The story of Brazil's evolution as an oil and gas nation is, in broad terms, one of success. That said, Brazil has faced challenges – and continues to do so – but the country has proved to be resilient in overcoming these challenges and exceeding expectations. Over the past 50 years, Brazil's oil and gas industry has been transformed: where once, it was barely able to meet domestic demand, today the country – and its national oil company – is internationally recognised and globally competitive, a transformation that has bought with it significant and diverse benefits.

The first commercial discoveries of oil were made in the State of Bahia following exploration activities in the 1930s. From the beginning, the concept of national ownership of Brazil's hydrocarbon sector was paramount and articulated clearly in the political messaging of the day: *O petróleo é nosso* – the oil is ours. At the time it was hoped that these discoveries would enable Brazil to become an international force within the growing global market for oil. There was a recognition that these newly discovered reserves offered the country the opportunity to change the economic and social landscape of the country. To achieve this vision – and to ensure that *O petróleo é nosso* would be more than just a slogan - the government founded Petrobras, the national oil company

Education and Training for the Oil and Gas Industry: The Evolution of Four Energy Nations.
http://dx.doi.org/10.1016/B978-0-12-800974-1.00003-9

of Brazil, in 1953. Since that time, Petrobras has played a leading role in growing Brazil's hydrocarbon sector and establishing the country on the international stage. Today Petrobras is recognised as being one of the biggest energy companies in the world, ranking third in 2010. For a national oil company to have risen to such heights is remarkable. Its reputation for tackling technologically complex operations distinguishes it from many other state-owned oil companies. Underpinning Petrobras's rise has been its ability to educate and train Brazilian citizens to a high standard. With this in mind, the story of Petrobras is central to the narrative around workforce development within Brazil's oil and gas sector. This case study will consider how Petrobras has achieved such sustained success and will explore the role of government, the effectiveness of local content policy and legislation and the role of other international players in building what one could argue is a world-beating oil and gas workforce.

COUNTRY FACTS

© *Shutterstock*

- Oil was first found in Bahia in 1930. In 1968, Brazil developed its first offshore oil project and 1974 saw the discovery of the largest Brazilian oil basin – Bacia de Campos in Rio de Janeiro.
- Brazil is the 10th largest producer of oil and gas in the world.[1]

---

1. US Energy Information Administration – http://www.eia.gov/countries/cab.cfm?fips=br.

- As of 2014, Brazil's total proven oil reserves were 13.2 billion barrels according to independent estimates. The Agência Nacional do Petróleo, Brazil's oil regulator, put the figure higher at over 15 billion barrels.[2]
- In 2013, crude oil production from the pre-salt layer was 303,000 bbl/d. This accounted for 15% of total production, a significant increase from when the pre-salt fields were first discovered. Investments estimated at US$53 billion will see pre-salt production increase to 18% of total production by 2015.[3]
- Domestic consumption of oil is higher than domestic production. The US Energy Information Administration predicts that domestic consumption will continue to be greater than production throughout 2015.
- Sustained economic growth has increased Brazil's consumption of energy putting it eighth on the list of global consumers.[4]
- Estimates released in 2014 predict that Brazil will produce 4.0 million barrels per day of oil by 2020. This will ensure that Brazil becomes a net exporter of oil with export levels predicted to reach 1.5–2 million barrels per day by 2022.[5]
- Petrobras expect that their contribution will be in the area of 2.9 million barrels per day between 2013 and 2020. This will rise to 3.7 million barrels per day between 2020 and 2030.[6]
- In 2012, petroleum and crude oil accounted for 8% of Brazil's exports. China was the largest importer of Brazilian crude oil in 2013, accounting for 30%. In recent years, United States imports from Brazil have decreased due to domestic production increases in the US.
- In 2013, oil and gas contributed around 12% of GDP. If the industry expands as planned, this figure should rise to around 20% by 2020.[7]
- Petrobras employed 80,908 people in 2014 within Brazil, overseas and across its subsidiaries.[8]
- The oil and gas sector currently employs 450,000 professionals. This number is estimated to increase to 2 million by 2020.
- On the World Bank Group's ease of doing business index, Brazil ranks 120 (for 2015) out of 189, which is an increase of three places from the previous year.

---

2. Ibid.

3. http://thebrazilbusiness.com/article/oil-industry-in-brazil.

4. U.S. Energy Information Administration – http://www.eia.gov/countries/cab.cfm?fips=br.

5. Platts, "Brazil to Export Up to 2 million b/d by 2018–2020: ANP", (September 15, 2014), http://www.platts.com/latest-news/oil/riodejaneiro/brazil-to-export-up-to-2-million-bd-by-2018-2020-21231640.

6. Petrobras, *2030 Strategic Plan and 2014–2018 Business and Management Plan*, http://www.investidorpetrobras.com.br/en/business-management-plan/business-management-plan/ano/2014.htm.

7. The Brazilian Oil and Gas Industry, 2013, pwc – http://www.pwc.com.br/pt/publicacoes/setores-atividade/assets/oil-gas/oeg-tsp-13.pdf.

8. Evolution of Petrobras System's Staff – http://www.petrobras.com.br/en/about-us/careers/.

- Brazil is the fifth most populated country in the world. In 2014, the population numbered 204,028,000 (with a population growth rate of 0.9% annually between 2010 and 2015).
- Brazil's elderly population will triple over the next four decades, from less than 20 million in 2010 to about 65 million in 2050 – this dramatic demographic change threatens Brazil's economic success.[9]
- According to the United Nations, Brazil's total labour force was 106,169,639 in 2014.
- In 2012, 5.5% of the labour force was unemployed.[10]
- 21.9% of the labour force was employed in the industrial sector in 2012.[11]

Oil drilling rig against panorama of Rio de Janeiro in Brazil.

A HISTORY OF BRAZIL'S OIL AND GAS INDUSTRY

The importance of the discovery of oil in Bahia in the 1930s is illustrated by the rapid actions of government to bring hydrocarbon resources under national ownership. The government saw these first oil discoveries as a strategic resource for national development and the slogan *O petróleo é nosso – the oil is ours –* encapsulated this. The sentiment demonstrated the belief that oil was for national development and was not something to be exploited by international companies seeking a quick profit. The government of Brazil was wary of the intentions of international oil companies having seen the emergence of an energy sector in Mexico within which the state had little to no control (until, that is, the establishment of Pemex, the national oil company of Mexico). The concept of

---

9. *Growing Old in an Older Brazil: Implications of Population Ageing on Growth, Poverty, Public Finance and Service Delivery* (2011) World Bank.
10. UN Data https://data.un.org/CountryProfile.aspx?crName=BRAZIL.
11. Ibid.

oil nationalism was made tangible in 1938 by then President Getúlio Vargas when he established the National Petroleum Council (CNP, O Conselho Nacional do Petróleo) as a mechanism to bring Brazil's oil industry firmly under state control. It should be noted that this is the same year the Mexican government enacted the expropriation decree that nationalised the assets of IOCs operating in Mexico. In Brazil's case, the establishment of CNP was not the precursor to the expulsion of IOCs from Brazil but rather was a first step towards the gradual nationalisation of the industry.

In 1953, Getúlio Vargas enacted law 2.004, allowing the creation of Petrobras, the state-owned oil and gas company. Petrobras would be operated under the authority of CNP and would assume responsibility for all oil and gas exploration and production activities as a means of ensuring that the economic benefits to Brazil of hydrocarbon discoveries would be maximised. As a means of protecting Brazil's national interests, the establishment of Petrobras was bold and, in many ways, effective. However, the economic climate created serious challenges. In the 1960s, Brazil was unable to meet its own domestic demand for oil and gas, making the country reliant on imports. This was in part due to the inability of the fledgling national oil company to achieve desired rates of production from existing fields. This had a profound impact on the Brazilian economy, particularly during successive oil crises that saw the price of oil tumble. The connection between oil and economy could be seen in the dynamics of Brazil's wider economic performance. In the late 1960s and early 1970s, Brazil's economy grew at a rate of 10% annually – the fastest growth in the world at the time – but the oil crisis in 1974 caused this figure to fall by half. Brazil's reliance on oil imports caused foreign debt to rise and left the country in a position where its continued economic growth was threatened. In an attempt to secure Brazil's energy future and isolate it from fluctuations in the oil price, the government established Braspetro, the international arm of Petrobras. Braspetro explored for oil in the Middle East, North Africa and Colombia during the 1970s. Braspetro discovered the Majnoon field in Iraq but it was largely unsuccessful and the company did not manage to secure the oil and gas deposits it was hoping to.

It was decided that Petrobras would focus on increasing production levels within Brazil so that domestic demand could be met. This meant Petrobras would need to form strategic alliances with IOCs in order to bolster low production levels. In 1973, the then President of Petrobras, Ernesto Geisel (who would go on to take office as the 29th President of Brazil less than a year later) allowed the state-owned company to enter into contracts with IOCs. Throughout this period, Petrobras's monopoly of the sector was never complete, and a number of oil companies operated in Brazil including BP, Texaco, Exxon, ELF, Conoco, Marathon and Total. These IOCs were joined by a handful of national companies also looking to benefit from Brazilian oil production opportunities. Amongst these were Azevedo Travassos, Camargo Correa and Paulipetro. Petrobras' monopoly of the oil sector was enshrined in the 1967 Constitution, but the practicalities of this monopoly were complex. Firstly, the constitution of 1967 did not detail how the monopoly was to be exercised and implemented. Secondly, the reality of Brazil's oil sector during the 1970s – and leading up to the 1988 Constitution – was quite different from what the 1967 Constitution had in mind: Petrobras was not the sole oil company operating within the country. While the envisioned monopoly may have not been actualised, it is true that Petrobras enjoyed some degree of dominance over the oil sector.

It would not be until the introduction of an amended constitution in 1988 that more restrictive measures would limit international oil companies from participating in exploration and production activities in Brazil. However, the limits put on international oil companies would not last for long and reforms would soon see state control come to an end. Changes in the Brazilian economy saw Law 9.478 (known as the Petroleum Law) established, allowing foreign competition in the oil and gas sector. Forty years of notional state dominance came to an end in 1997 and international oil companies were officially invited to bid on concessions in Brazil. Under new legislation other agencies were created to manage the oil sector in light of foreign involvement, including the National Petroleum Agency (Agência Nacional do Petróleo, ANP), who were made responsible for the regulation and supervision of petroleum activities, and the National Council of Energy Policies, who developed policy related to energy. Petrobras would, under this legislation, have to compete with international operators for concessions under the governance of ANP. At the same time, Petrobras began to reshape its business, looking beyond Brazil's borders for commercial opportunities and building on the company's reputation as a leader in technically challenging offshore extraction. As a result, Petrobras successfully signed agreements with a number of South American countries before looking further afield. Additionally the early 1990s was a period of managerial changes within the company; management was decentralised from the government and Petrobras was allowed to act and operate as a private company. Savvy, business-minded individuals, coupled with decades of technical expertise, enabled Petrobras to become an international corporate entity with operations across Latin America and the rest of the world.

The transition from being a highly centralised state-run organisation to a decentralised, dynamic company was well timed and represented an important stage in the evolution of Brazil's oil and gas industry. It gave Petrobras the freedom to pursue business interests in a way a private corporate entity might, making it more efficient and focused on cost-effective production. Yet its strong links to the government ensured that Petrobras still operated within a context of supporting the wider economic goals of the country. Furthermore, under the guidance of President Geisel, Brazil was able to adopt a pragmatic approach to dealing with other countries. Under his presidency, Geisel ensured Brazil remained on good terms with the USA while building closer ties with China and Angola, amongst other countries. This yielded economic opportunities for Brazil in Africa and Asia and contributed towards Petrobras' strategic, global approach to oil production. Further to this, Geisel began a slow process of democratisation that removed tight military regulation of the political scene.

This enabled Brazil as a country to move forward with an economic liberalisation agenda and gave it more of a presence on the international stage. The introduction of a wider democratic process meant that Petrobras, in time, would be able to distance itself from the machinery of government and pursue business interests that would, in the long run, be fruitful for Brazil's economy.

Since 2000, Petrobras has expanded its interests even further afield. 2003 saw the company acquire Perez Companc Energia, an Argentine energy operator. This further opened up key South American markets to Petrobras, including Peru, Paraguay and Bolivia. 2005 saw Petrobras enter the Japanese market through a contract with Nippon Alcohol Hanbai. This deal secured Brazilian ethanol exports to Japan. 2006 and 2007 were important years for Petrobras; a new field was discovered – the Tupi oil field in the Santos Basin – which represented the Western hemisphere's largest oil discovery of the last 30 years.[12] The Tupi field is estimated to contain 7500 million recoverable barrels of oil and is located in the pre-salt layer. The Jupiter field (discovered at a similar time) is estimated to equal the Lula oil field in size. These discoveries were hailed as a 'second independence for Brazil', according to the former president, Luiz Inácio Lula da Silva. Pre-salt formations are common in Brazil and typically expensive to develop. In pre-salt formations, the oil and gas lies below a 2000 m layer of salt which sits below a 2000 m layer of rock under the Atlantic, which is itself 2000–3000 m deep. In recent years, Petrobras has successfully drilled fields located in the pre-salt layer. Pre-salt extraction accounts for 15% of Brazil's production (2013), a significantly higher figure than preceding years when Brazil first discovered and began production on pre-salt fields. Pre-salt operations have contributed significantly towards Brazil becoming one of the top 10 producers in the world:

> The deposits discovered in the pre-salt sedimentary basins of Campos and Santos offshore Sao Paulo and Rio de Janeiro (in the Southeast region), should provide the country with an income of US$28 billion from oil exports in 2020. According to the study, export volumes of 600 thousand barrels per day are expected by the end of this decade. If these figures are confirmed, this will be an increase of 73% in relation to 2010, equivalent to US$16.1 billion.[13]

At the end of 2008, Petrobras had an annual income (net) of US$130 billion, investments of US$32.3 billion, production nearing 2 million bbl/d, 422,000 bbl/d, 13,174 productive wells, 16 refineries and 112 platforms.[14] Since beginning operations, Petrobras has enjoyed considerable success and is viewed as a hugely effective enterprise amongst analysts and global development agencies

---

12. Jeff Flick, *Petrobras Pumps First Crude from Massive Tupi Field Offshore Brazil* (Rigzone, May 1, 2009). Retrieved January 8, 2010.

13. Brazil, "Oil & Gas Industry in Brazil: a brief history and legal framework", Lais Palazzo Almada and Virginia Parente (2013).

14. Andrew D. Fishman, *Petroleum in Brazil: Petrobras, Petro-Sal, Legislative Changes & the Role of Foreign Investment.*

when compared to other state-run companies around the world.[15] Although there are a significant number of operators in Brazil's oil and gas sector, Petrobras was responsible for 91.2% of crude oil produced in the country during 2010 indicating its continued dominance of the sector in spite of the presence of IOCs.[16] In its planned investments for the 2011–2015 period, the company set aside US$127.5 billion for exploration and production, US$70.6 billion for refining, US$13.2 billion for gas and energy projects, US$3.8 billion for petrochemical activities, US$4.1 billion for biofuels, US$3.1 billion for distribution and US$2.4 billion for other corporate activities. In total, Petrobras invested US$224.7 billion over this 4-year period. The scale and scope of these investments is testament to the size and ambition of the company.

Petrobras is by far the biggest player in Brazil's oil industry, but other international companies and national entities merit attention. Since 2011, private companies native to Brazil have begun exploration activities in the country. Companies such as OGX and HRT O&G have invested huge amounts in beginning production activities within Brazil's borders. In 2011, Brazil's oil and gas industry looked like this[17]:

Company	Number of Rigs (2011)
Petrobras	68
Shell	1
Chevron	2
BG	
Statoil	2
Frade Japao	
Devon	
ONGC	1
SK	
Repsol YPF	
Petrogal	3
OGX	8
HRT O&G	3
Imetame	2
Total	98

Between 2011 and 2015, private companies invested some US$36 billion in exploration and production. BG Group alone plans to invest a total of US$30 billion by 2020 (with a further US$1.5 billion invested in research and development in country by 2025). Since 2011, China's interest in the Brazilian oil and gas industry has rapidly increased, with well over US$10 billion being invested by various Chinese petrochemical companies.[18]

---

15. Mahmood A. Ayub and Sven O. Hegstad, "Management of Public Industrial Enterprises", *The World Bank Research Observer 2.1* (1987): 82.
16. ANP – *National Petroleum Agency, Bulletin of the Oil & Gas Production* (November, 2010).
17. All figures taken from the *Brazilian Institute for Oil, Gas and Biofuels IBP*.
18. *The Brazilian Oil and Gas Sector* (Brazil: Swiss Business Hub, September, 2011).

The government has established a legal framework governing hydrocarbons and their extraction and these laws form the basis for international participation. Amongst these laws are:

- The Petroleum Law (Law 9.478): passed in 1997, the historic law that allowed international oil companies to participate in Brazil's oil and gas sector. This law reformed the oil industry and also established the National Agency of Petroleum – ANP – who is also responsible for developing local content regulation. The law also established the National Council for Energy Policy.[19]
- Law 12.351: Passed in 2010 and known as the production sharing agreement law (PSA law). This law provides a production sharing model that emulates the models used by many other national oil companies and involves a mandated partnership between the NOC and an international oil companies for the production of a particular concession. IOCs are entitled to a portion of production allowing them to recover investment costs. Profits are then shared between the state and the IOC.
- PSAs in Brazil will also govern the exploration and production of pre-salt fields with all contracts signed by the Pre-sal Petroleo S.A. (PPSA), the state-owned company that authorises the bidding processes for pre-salt concessions. Law 12.304/2010 established Pre-sal Petroleo S.A. to ensure the profitable pre-salt layer is exploited according to Brazil's regulations and that this activity generates profits that benefit the nation.

The story of Brazil's hydrocarbon sector is one of gradual nationalisation and a focus on national ownership of the country's oil and gas deposits. Having seen the troubling dynamics between other Latin American nations and international oil companies, Brazil was quick to set in motion a clear policy on national ownership whilst maintaining a degree of international participation in its hydrocarbons sector. Despite Petrobras' monopoly of the sector being enshrined in the constitution, internal pressures have created an industry where risk is shared between the state-owned Petrobras and a number of multinational oil companies. Brazil's transition to a democratic, modern country with a dynamic economy has seen international operators playing an increasingly bigger role in exploration and production (albeit that Petrobras remains as the dominant force). Petrobras' international achievements, measured by its global assets, reflect the country's historical need to look outwards, beyond its borders, for economic growth. For a national oil company to establish itself as an international force within the highly competitive hydrocarbon sector is testament to Brazil's willingness to build international partnerships as well as its ability to create a knowledge and skills base that can compete with the rest of the world.

---

19. Andrew D. Fishman, *Petroleum in Brazil: Petrobras, Petro-Sal, Legislative Changes & the Role of Foreign Investment.*

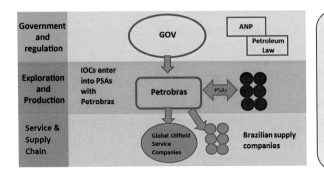

Brazil's Oil and Gas Sector.

A HISTORY OF EDUCATION AND TRAINING IN BRAZIL

The history of Brazil's education system is inextricably linked to the progress it has made in becoming one of the largest economies in the world. Economic success has been underpinned by a recognition of the vital part that a coherent and effective education and training system plays in building a competent workforce. The focus on achieving universal access to education across the board, from primary to tertiary, has also been a key feature of this story. One of the historic strengths of the Brazilian approach has been to build the quality and availability of higher education institutions whilst, simultaneously, creating a system of technical and vocational study that can respond to the needs of industry and develop the skilled artisans and technicians needed to grow the economy. This focus on both the academic and the technical has its roots in the late eighteenth century with the establishment of the Escola Politecnica (polytechnic school). The Portuguese authorities began issuing decrees for the creation of engineering schools, which led to the establishment of a number of military engineering institutes. It was during this period that the Escola Politecnica da Universidade Federal do Rio de Janeiro (Polytechnic School of the Federal University of Rio de Janeiro) was opened making it one of the oldest schools of engineering in the world.

Brazil's economy underwent a transformation soon after World War II, leading to the government focussing on improving the availability and quality of education. In this period, Brazil was able to increase the provision of university education. Unfortunately, efforts made to improve tertiary level education were not mirrored within the primary and secondary streams. Brazil quickly fell behind other developing and developed nations in terms of its provision of K-12 education. Issues with low levels of enrolment and a lack of access to schools across Brazil's states hampered progress. Acknowledging that primary and secondary education was crucial to capacity building, the government began to invest in improving the accessibility and quality of schooling.

Over the last 20 years, Brazil has fought a battle to not only increase attendance at schools, but also to retain students and see them graduate from secondary education. Many students today are the first in their families to finish high school. Brazil now enjoys high attendance levels within primary and secondary school with figures showing an increase in the number of children aged 7–14 attending school and reaching nearly 100% in 2000. Despite increasing the provision and quality of basic education, Brazil came 53rd in the PISA international study of education systems in 2012. In 2013, reports emerged showing that Brazil was one of the countries where employers found it most difficult to find personnel with the right skills and talents to fulfil positions. According to the Talent Shortage Survey, 68% of businesses in Brazil reported talent shortage as a major factor behind filling jobs.[20] But this trend is far older than the report. Rapidly increasing industrial activity has created pressures on skilled talent within the workforce and this has put a spotlight on the need to educate and train vocationally skilled workers. Additionally, Brazil has planned large-scale infrastructure projects, including roads, airports and stadia. The 2016 Olympics and the 2014 World Cup were also factors that highlighted the ongoing pressures on Brazil's technical capacity. The pace of economic growth has quickened to a point where the education sector struggles to produce the required manpower to meet skills demands.

The battle to improve skills development in Brazil has been long and there is broad recognition that the education system must increase the number and quality of its technically skilled graduates. Despite the Brazilian government taking advantage of the global financial crisis to encourage technical expertise into the country from Europe, the challenges of developing a skilled workforce cannot be met through immigration alone. The problem is systemic and must be addressed through a comprehensive education strategy that starts in basic education and extends through technical and vocational and tertiary study. However, to suggest that Brazil has failed in these efforts would be misleading. There has certainly been an issue with quantity – with simply not enough young people graduating with the skills required to fulfil demand – but there have also been notable successes, particularly in relation to technical and vocational study.

Technical and Vocational Education and Training

Since the early 2000s, Brazil has made strides to address technical skills development through a series of projects aimed at building the strength of the technical and vocational education and training (TVET) system. Most significantly, education funding rose from US$385 million to US$3.8 billion

---

20. *Talent Shortage Survey*, 2013.

within the space of 10 years.[21] Vocational education (widely known in Brazil as 'professional education') is delivered through a network of 500 technical institutes. There is also legal framework in place for the provision of TVET in the form of the National Education Guidelines and Framework (Lei de Diretrizes e Bases da Educação). This legal framework sets out the responsibilities that different parts of the education system have for skills development and focuses on the preparation of students – within the secondary and tertiary stream – for employment.

Although recent efforts at TVET reform are an indication of the challenges that remain across Brazil, the country does offer a model of TVET that has been in operation for many years and that has reaped significant benefits for employees, employers and the economy. This system warrants analysis. The SENAI (Serviço Nacional de Aprendizagem Industrial or National Service for Industrial Training) system was established by industry leaders and President Getúlio Vargas in 1942 as a network of professional institutions at the secondary level directly designed to improve local capacity and provide vocational and technical training to people across Brazil. SENAI operates across all of Brazil's states and has, over decades of operation, made a significant contribution to workforce development. SENAI has forged ties with the Ministry of Education and the Ministry of Labour, invested in expanding the number of courses offered in its fixed and mobile units numbering over 800 across every state, improved the quality of its training by sending staff abroad to undergo teacher training and has successfully built partnerships with German, Canadian, Japanese, French, US and Italian institutions. The SENAI initiative offers free courses to Brazilian citizens in a number of industrial areas. Alongside its offer, SENAI is able to provide training to students who may not be able to make their way to physical institutions. This is achieved through a fleet of mobile units that have created education and training opportunities to those looking to improve their skills or access new trades/industries.

SENAI has achieved a great deal in nearly 70 years of operation:

- 55 million qualified professionals between 1942 and 2011;
- 2.5 million annual enrolments in introductory courses, learning skills, mid-level technician, technologist, senior, graduate and professional development;
- 3300 municipalities served;
- 140,000 technicians and laboratory services performed in 2011;
- 48 international partnerships with 30 countries;
- 10 vocational training centres in operation and 6 under negotiation.[22]

---

21. Liz Dwyer, *Can Brazil Teach Us How to Get Over Out Vocational Education Snobbery?* (July 11, 2011).
22. Official SENAI Website – http://www.portaldaindustria.com.br/senai/institucional/2012/03/1,1774/atuacao.html.

SENAI has played an important role in strengthening Brazil's technical workforce as evidenced by the above figures. This success is multiplied by other organisations belonging to the Federal Centres for Technological Education (CEFET) and so-called 'System S'. These organisations are financed by levies paid by companies who belong to the system (with the levy set as a 1% business tax). The system is overseen by the Ministry of Education and the Ministry of Labour. Alongside SENAI, other participating organisations belonging to the system are the following:

- SENAC – National Service for Commercial Training (Serviço Nacional de Aprendizagem Comercial);
- SESC – National Service for Business Training (Serviço Social do Comércio);
- SESI – Social Service for the Industry (Serviço Social da Indústria);
- SENAR – National Service for Rural Training (Serviço Nacional de Aprendizagem Rural);
- SENAT – National Service for Transport Training (Serviço Nacional de Aprendizagem do Transporte);
- SEST – Social Service for the Transport sector (Serviço Social do Transporte);
- SEBRAE – Brazilian Services for Assistance to Micro and Small Companies (Serviço Brasileiro de Apoio às Micro e Pequenas Empresas); and
- SESCOOP – National Service for Cooperative Learning (Serviço Nacional de Aprendizagem do Cooperativismo).[23]

Through the establishment of the 'System S' – and related initiatives – Brazil has made significant strides towards the provision of an effective TVET system. There is a clear qualifications framework and a structured pathway for student progression. Furthermore, the education system is strongly aligned with industry and this gives students the opportunity to gain on-the-job experience. However, there has been, and remains, a shortage of technical workers for Brazil's industrial sector, although this has less to do with the provision and quality of TVET in Brazil and more to do with the rapid expansion of Brazil's economy.

Tertiary Education

The number of tertiary education institutions in Brazil has increased throughout the twentieth century and evidence suggests a steady improvement in the quality of education provision. In the first half of the twentieth century, Brazil's tertiary education institutions were largely attended by the country's wealthier citizens. With the collapse of the military-controlled government and the reintroduction of democracy, Brazil was able to make gains in opening access to tertiary education to a larger number of people from more diverse socioeconomic and ethnic backgrounds. As part of the transition to

---

23. http://www.unevoc.unesco.org/go.php?q=World+TVET+Database&ct=BRA.

a democratic state, the new 1988 Constitution laid out the rights of citizens, which included the right to education. One of the most important features of the new constitution was the ability it gave government to allocate public funds into schools and universities, whether public or private. In 1996, another important development allowed private universities to operate as for-profit entities. This liberalisation of tertiary education is now an important part of the education landscape in Brazil.

The Brazilian university system is, today, internationally recognised for its academic rigour and the quality of its graduates. Brazilian institutions appear within the top 200 best universities in the world. The 2012 University Ranking by Academic Performance (URAP) ranked The University of São Paulo 28th in the world. This is a significant achievement for a country with a relatively young comprehensive, universal education system. Furthermore, Brazilian universities rank high in Latin America, with the University of São Paulo, Universidade Estadual de Campinas (Unicamp), Universidade Federal do Rio de Janeiro, Universidade Estadual Paulista 'Júlio de Mesquita Filho', Universidade Federal de Minas Gerais and Universidade Federal do Rio Grande do Sul ranking second, third, fourth, ninth and tenth respectively on the list of top 10 universities in Latin America.

The university sector in Brazil is a mix of public and private institutions. Public universities are generally seen to offer the best level of education and, with tuition being free in most cases, competition to enter public universities is fierce. Despite significant investment, public universities complain that they are underfunded and thus unable to increase their staff numbers or invest in their facilities in order to enable them to allocate more places on degree programmes.

Private universities in Brazil do not enjoy the same reputation as their public counterparts. However, with private universities offering night classes, they are becoming a popular choice amongst the employed seeking to undertake further education. Private universities have been improving, and in 2009, the private education sector became the 10th largest sector in the Brazilian economy, demonstrating the viability of private education in Brazil as a business. Between 2000 and 2009, the number of institutions offering private education more than doubled from 1004 to 2069. Brazil currently has over 2600 public and private universities, and this figure is quickly growing.

Brazil has made impressive progress towards improving education at all levels. Since the 1940s, the country has recognised that the continued development and expansion of its fast-growing economy is directly related to the number of technically skilled workers graduating from the education system. Consequently primary and secondary education is now universally available and there has been an unprecedented level of students completing high school and going on to tertiary education. The government has successfully implemented a comprehensive system of TVET, and this system has a long arm, reaching most communities, both urban and rural, through its network of institutions and mobile learning units.

It is also important to note the following:

- There is a national qualifications framework and legislation addressing the provision of TVET. Strong policy has helped to increase the number and quality of courses provided, the quality of instruction and the applicability of the programmes
- Amongst TVET institutions there is strong alignment with the industry
- Brazil's model for TVET relies on both public and private organisations cooperating
- In 2012, Brazil's lower house of congress approved a National Education Plan that detailed objectives to increase the spending on education to 10% of GDP by 2020
- Brazil has a strong, internationally competitive university network. Graduates are academically mobile and degrees are recognised internationally

Oil platform at Guanabara Bay 2013. *(Attila JANDI/Shutterstock.com)*

THE STORY OF WORKFORCE AND SKILLS DEVELOPMENT FOR THE OIL AND GAS INDUSTRY

When oil was discovered in 1930 and the government acted to nationalise reserves, there was a realisation that the country would need to train and educate engineers, technicians and a host of technically skilled workers if national participation in the nascent sector was going to be achieved. The establishment of Petrobras was one of a number of steps to ensure that the country directly benefited from production and this placed the company at the centre of efforts towards workforce and skills development. As the country moved away from military rule and implemented a process of democratisation, efforts were made to increase the skills and knowledge of the workforce. Under the 'citizen constitution' of 1988, the right to education was established and the government was granted the ability to channel funds into education. This meant that oil revenues would be directly invested into education nationwide, providing citizens with free access to schools and colleges. Today Brazil's education system is recognised as one of the most successful amongst the so-called 'BRIC' countries. Its tertiary education system is able to compete internationally and develop a workforce that is globally competitive. The education sector is also a major political platform for Brazilian politicians and, considered within this historical context, it is unsurprising that the government plans to raise investment into the education sector to 10% of GDP by 2020. Bearing in mind the success that Petrobras has enjoyed both nationally and internationally, we can now analyse the factors that have enabled Brazil to build and develop its oil and gas workforce.

The Role of Government and Local Content

In spite of the pre-eminence of the state oil company (which one might assume would limit the need for local content policy), successive Brazilian governments have been careful to balance local content requirements across the industry and the wider supply chain, exploring models that are sustainable and realistic and then applying these models to the oil and gas sector in a way that works within a Brazilian context. Although the effective implementation of a local content policy is not, in itself, directly linked to workforce development, building local participation creates demand in the market and builds the need for skills. This, in turn, creates a context within which skills and workforce development have a clearly articulated aim and a compelling economic objective.

A great deal of effort was put into developing local content policy and economic growth strategies that would be fit for purpose in Brazil. On the whole Brazil's local content policy and legal framework has enjoyed some measure of success in promoting local participation and, in so doing, creating a demand for skilled labour.

Local content policy works best when implemented as part of a wider, macroeconomic structure and this is exactly how Brazil has approached the challenge of improving local participation. In effect, Brazil's local content policy strengthens demand directed to the domestic market (the goods and services industries that make up the supply chain of the oil and gas industry, such as construction and naval engineering). Through improving the demand for other industries the local content policy not only ensures direct local participation in the oil sector, but it also supports the growth of the wider economy by creating jobs in other industries and supporting home-grown companies. The long-term objectives have worked on two fronts: first, the diversification of the industrial sector; and second, the development of technology and research-driven sectors.

- In terms of the diversification of the industrial sector Brazil's local content requirement is aimed at raising participation in the domestic good and services industry in the oil and gas supply chain on a competitive basis. This policy is supported by the Program for the Mobilization of the National Industry of Oil and Natural Gas (PROMINP). PROMINP is led by Petrobras and the National Development Bank (BNDES). The PROMINP programme was established in 1993 and aims to maximise the involvement of domestic goods and services suppliers in Brazil's oil sector, in alignment with the local content policy. There are some exceptions and deepwater subsea equipment is still largely imported from outside Brazil. Yet gains have been made and local suppliers are capable of competing with their overseas counterparts and are able to meet the standards of international suppliers. Tools for remote operating vehicles and pipe pigging are now supplied through domestic companies, but Brazil has some way to go before it can provide the oil and gas industry with the complex equipment needed to develop challenging fields.

- Research and development for the supply of technologies and equipment is crucial to diversification of the industrial sector and to growth in the oil and gas industry. Without a strong network of research centres, domestic companies cannot compete with international goods and services suppliers. The federal government collaborated with the state government of Rio de Janeiro to incentivise companies to place their research and development centres in the Technology Park of Ilha do Fundao – Rio de Janeiro. This Park is situated close to CENPES, the Centro de Pesquisas Leopoldo Américo Miguez de Mello and enjoys close links to university research networks. The Technology Park has received more than US$260 million in government funding. The Park is also home to the research and development departments of Schlumberger, FMC Technologies, Baker Hughes, Halliburton, Tenaris Confab, Usiminas, GE, EMC2, Chemtech and BG Group. Significant investment in research centres such as this ensures that major international companies are attracted into the country.

The positive impact of Brazil's local content policy is largely due to the mechanisms by which the policy is enacted and made real. In Brazil's case, these mechanisms are Petrobras and PROMINP. Petrobras plays a crucial role in the implementation of local content policy and has an agreement with SABRAE a national association that supports small businesses. Through this agreement Petrobras is able to include small- and medium-sized business in the oil and gas supply chain. The agreement allows Petrobras to map the landscape of business opportunities and channel funding into the right areas so that training and support can be given to those companies most in need of it. In this way, Petrobras is able to support the capacity of local companies and they become realistic alternatives to international suppliers. The second tool of local content policy enactment is PROMINP. PROMINP identifies the challenges to implementing local content requirements and addresses these challenges through training and education. It is through PROMINP that the dual pillars of local content and local education and training become aligned (See 'A note on PROMINP' below for more on this initiative).

The funding of education and training is also tied up with operations within the oil and gas sector. As part of local content policy, the approach to spending and the sourcing of goods and labour is established during each licensing round and operators are required to invest 1% of gross revenue from each field being produced. This revenue is channelled into research and development projects relating to the oil and gas industry. Half of this must be directed towards Brazilian universities or ANP-accredited institutes.

As a result of the funding and investment that has poured into Brazil's tertiary education sector, overseas companies have opened factories and manufacturing plants in Brazil having been attracted by the strong research base that this investment has created. These companies are active in local research initiatives and technology transfer for deepwater equipment – important in the wake of large pre-salt discoveries. This exchange between foreign and local companies is hugely significant to developing a workforce with the right specialisations for pre-salt extraction and production.[24] It is also important to mention research initiatives into pre-salt extraction technologies as much of Brazil's future depends on exploiting the hydrocarbons found in these deep reserves. Operations in pre-salt fields are typically expensive and technologically complex. Without the long-term local policy objectives devoted to research and development, Brazil will not meet pre-salt extraction targets.

A Note on PROMINP

PROMINP (Mobilization of the National Industry of Oil and Natural Gas) was established by the federal government in 2003 and functions as a forum for stakeholders to discuss development initiatives. The idea is that PROMINP – run by the Ministry of Mines and Energy and Petrobras – promotes the growth of industries in the goods and services sector. PROMINP's governing body includes the Minister of Mining and Energy, the President and Services Director of Petrobras, the Minister of Development, Industry and Trade Ministry, the President of the Brazilian Petroleum Institute, the President of ONIP (Organização Nacional da Industria do Petróleo) and the President of BNDES (National Bank for Social and Economic Development).

Under the initiative, industries work with the oil and gas sector to provide technology and other essential goods and services that the sector depends on. Since its creation, PROMINP has nurtured many of these essential industries

24. http://en.mercopress.com/2013/12/26/brazil-conditions-any-future-oil-pre-salt-bidding-to-local-industry-capacity-to-supply-equipment-and-services.

and domestic companies are now able to manufacture goods that were once imported from overseas. The growth of the domestic production chain has led to significant job creation and by 2020, PROMINP expect 38 platforms, 28 rigs, 146 support vessels and 88 large vessels to be built by the domestic supply chain as a direct result of PROMINP's intervention.

Initially PROMINP carried out a survey to evaluate national competitiveness so that local supplier capability could be analysed and problems identified. Supplier capability was compared to expected demand from the oil and gas sector looking over a 20-year period. This analysis was then used to create an action plan covering capability gaps.[25]

Although PROMINP is a national initiative, it is very visible at the local level. PROMINP forums exist across a number of Brazilian states, including Alagoas, Amazonas, Bahia, Ceará, Espírito Santo, Maranhão, Minas Gerais, Paraná, Pernambuco, Rio de Janeiro, Rio Grande do Norte, Rio Grande do Sul, São Paulo and Sergipe. These regional fora allow the PROMINP initiative to evaluate the needs of supply chain industries at the local level and make local arrangements. Working at the local level gives small business the opportunity to compete for development opportunities and fosters their growth in a way that could not be achieved if PROMINP did not seek to understand what was happening at the state level. PROMINP can synchronise state agendas with the national agenda for business development based on the growth of the oil and gas industry. This decentralised approach is highly effective.

PROMINP serves as one of the tools for local policy strategy in Brazil. Simply put, local content sets out terms and PROMINP makes capacity building a realistic goal for international oil companies. Local content legislation requires international oil companies to assess where they are sourcing labour, assets, systems, subsystems and services. PROMINP assists local suppliers to develop their businesses and build the capacity of their staff so that they can offer an attractive proposition to the industry. PROMINP is one example of how the Brazilian private sector is working with the public sector (at the state and national levels) to create real economic growth based on realistic – though optimistic – goals. According to the World Bank PROMINP has significantly raised the participation of local suppliers in the oil and gas sector from 57% in 2003 to over 75% in 2009, resulting in 755,000 additional jobs for Brazilians.[26] PROMINP has trained approximately 88,000 professionals since its inception with a further 200,000 professionals being trained for 185 technical positions.[27]

---

25. World Bank, *Local Content Policies in the Oil and Gas Sector.*
26. Ibid.
27. Official figures from Petrobras: http://www.petrobras.com.br/en/our-activities/performance-areas/oil-and-gas-exploration-and-production/pre-salt/.

Construction of a Petrobras oil refinery 2013. *(A.RICARDO/Shutterstock.com)*

Petrobras' Approach to Education and Training

Petrobras could be considered the most important organisation in the implementation of local content policy in Brazil. Since its inception in 1953, Petrobras has driven workforce development, education, training and has promoted job opportunities for local people within the oil and gas industry. Early on Petrobras realised that if the oil and gas industry was going to benefit local people, then the country would need technically skilled operators and engineers to assume important roles in the industry. Petrobras faced the same problem many emerging energy nations face today – how to build and maintain a technically competent workforce for its new oil and gas industry? The company began by investing in tertiary education initiatives, which included the Petrobras University, established in 1955 to train specialists in oil exploration (more of which later). Since then, Petrobras has engaged other parts of the education and training ecosystem and has worked with the private sector and federal government to create a web of workforce development programmes, working alongside, and as part of, the national education system. It is within this context we can look at Petrobras' approach to developing local capacity.

A key partner for Petrobras in achieving its workforce development objectives has been the SENAI initiative. SENAI has increased the competitiveness of Brazilian industries through promoting technical and vocational education and sits within the S-system, an extensive out-of-school training system that has been providing advanced skills training to Brazilian citizens since the 1940s (and mentioned earlier in this case study). The S-system has been critical to the skills development activities that have underpinned Brazil's industrial boom and now offers a considerable range of programmes across energy, agriculture, commerce and small business management. SENAI enjoys a high degree of success with centres located in all of the main oil- and gas-producing regions, giving it an important role in skills development for the oil and gas industry. SENAI and Petrobras have a long history of partnership around skills development and today offer free courses in welding, mechanics, electricity and industrial instrumentation. Further to this Petrobras and SENAI partner together to promote youth education – the 2-year long programme known as the Petrobras Young Apprentice Program[28] seeks to create opportunities for young people from socially vulnerable backgrounds to work in the oil and gas industry through a regime of training. The structure of this training is as follows:

- Basic training for 4 months in identity and citizenship. The candidates will also use this time to better understand Petrobras' roles, responsibilities and operations.
- SENAI provides a 9-month professional qualification course that builds students' knowledge and skills in the relevant industry area.
- Petrobras then take candidates and trains them on the job as apprentices with professional supervision from local managers of Petrobras units.[29]

Candidates receive a minimum wage for the time they are enrolled on the course, compensation for food and travel expenses and vacation in accordance with the legal framework for apprenticeships. By 2015, figures showed that 555 young people had undergone or were undergoing training as part of the Young Apprentice Program.

Amongst the other successful programmes accredited to the Petrobras–SENAI partnership is the Gas Technology Centre, now known as the Centro de Tecnologias do Gas & Energias Renovaveis (Gas Technology Centre and Renewable Energy), or CTGAS-ER. CTGAS-ER, since its genesis in 1999, has trained around 40,000 students at the secondary, higher and postgraduate levels in a number of courses around renewable energy and natural gas technology. Through a range of professional training courses, candidates are able to undergo vocational instruction and eventually specialise in a specific technical area. In total, the Centre dedicates 78,000 hours annually to providing technological services and research to clients.[30] This is in keeping with SENAI's overall success at combining vocational education with technical research – SENAI has the largest private network of laboratories in Brazil through which it supports the development of innovation for the sector.

SENAI – and more generally the S-system – though highly successful, has weaknesses. Commentators note a lack of tertiary level links to the system. This disconnect has limited the effectiveness of training for candidates. Innovation has also been hampered by this lack of interaction and this is now impacting on businesses that rely on knowledge transfer and skills training. However, SENAI and the S-system have been models of best practice for other Latin American

28. http://www.petrobras.com.br/pt/quem-somos/carreiras/oportunidades-de-qualificacao/jovem-aprendiz/.
29. http://www.petrobras.com.br/pt/quem-somos/carreiras/oportunidades-de-qualificacao/jovem-aprendiz/.
30. Official figures taken from CTGAS-ER Website – http://www.ctgas.com.br/index.php/institucional/apresentacao.

countries looking to replicate Brazil's success in the area of workforce development. One of the strongest aspects of the SENAI model is its ability to adapt education and training provision to the changing needs of the market. During the economic crisis of the 1980s, SENAI adopted new practices in the wake of economic shifts. At this time, SENAI invested in technology and in developing specialist staff. It worked more closely with companies, increased on-the-job training, set up centres for research and development and populated these with specialists and the latest equipment. Since its inception some 39 million enrolments have been recorded. There are over 2 million enrolments annually. 744 operational units now offer approximately 1800 courses and 80,000 technical services annually to Brazilian companies.

At the vocational and skills development level, FIRJAN (Federation of Industries of the State of Rio de Janeiro) works closely with SENAI, PROMINP and, by extension, Petrobras, to guide the decision-making processes in the oil and gas sector in Rio de Janeiro. This includes providing the latest technology and equipment for training and generally assisting companies and improving their competitiveness. In 2013, Petrobras came to an agreement with FIRJAN and ASET (Aberdeen Skills and Enterprise Training Ltd). This agreement was for an investment of US$83.6 million to fund the development and installation of new simulators to be used for a range of training programmes. The main customer and supporter of the new simulator deployment was Petrobras. As part of the plan, Petrobras wanted 924 employees to be trained for a period of 7 years. For the company, this agreement represented a huge cost-benefit, allowing Petrobras to train people locally. Until then all the company's platform operators were trained overseas in Aberdeen. Petrobras is not the only company to train employees using new equipment. The installation will enable other companies from across Latin America to take advantage of training that is closer to home. The ballast simulator is installed in the Offshore Training Center (NTO) Engineer Nelson Stavale Malheiro. The unit simulates – down to exact details – the cockpit of an oil rig. Weighing 12 tonnes, it is supported on an electric jack driven by two motors and mimics the movements and noises felt and heard on a real platform. This simulator is being used to teach operators about stability issues, emergency management and major emergencies related to the types of rigs used by Petrobras.

FIRJAN statistics show that in the first half of 2014 companies operating under the FIRJAN umbrella were able to successfully train 40,000 professionals for the oil and gas industry (one of Rio de Janeiro's most important industries responsible for 72% of Brazil's overall oil production).[31]

This network of private and public entities working together to create and deliver national education strategies is an important part of why Brazil has been so successful at training specialists for technical occupations. The ability of

---

31. Official figures taken from FIRJAN Website: http://www.firjan.org.br/petroleoegas/institucional/institucional.htm.

Petrobras to collaborate with others – and to clearly articulate its skills and workforce needs – coupled with proactive approaches to the funding of research and technology have created a sound basis for skills and workforce development at the technical and vocational level that has rightly been seen as an exemplar to other countries wanting to achieve similar objectives. The partnership between Petrobras and SENAI is particularly crucial and the impact on skills development within the oil and gas sector has been profound. One may argue that the emergence of Petrobras as a company of international standing is built on the foundation of these partnerships.

Petrobras and Higher Education

Whilst Brazil's approach to skills development is to be applauded, the need for higher-level expertise has by no means been neglected. Petrobras has played a significant role in the development of engineers both within it's own initiatives and through partnerships with overseas agencies. This has been an important feature of Brazil's education landscape for the oil industry and has played a key role in strengthening national capacity. In 1955, Petrobras created an education centre where courses were offered to train and specialise people for oil and gas exploration and production. This institution was called the Center for Education and Research of Oil (CENAP) and was a partnership with the University of Bahia enabling drilling and oil production programmes to be rolled out as part of wider Petrobras education and training initiatives. In 1956, CENAP introduced the Course of Petroleum Engineering. This course was the first to be introduced as part of company's internal education system and it successfully trains petroleum engineers to this day. The number of petroleum engineers who have graduated from this course since 1958 number over 2700.

Candidates are selected from amongst the best performers in nationwide tests. Those who pass are then invited to undergo psychological and social evaluations to determine their suitability for an oil and gas career. Upon acceptance students spend the first 2 weeks developing an understanding of Petrobras and its activities. Students will also spend 47 weeks learning what is known as the 'Technical Dimension', the technical aspects to exploration and production. The Technical Dimension is split up into five periods covering five disciplines. Each period lasts for 1000 hours (on average). Failure to complete or pass a period results in dismissal from Petrobras and the programme.[32] This rigorous process of selection and training filters out candidates not suited to work in the oil and gas sector. This has two major advantages: Firstly, Petrobras saves money and prevents losses incurred from recruiting the wrong people and, secondly, the high standards of the company help to give it a sense of exclusivity, raising perceptions around employment in the oil and gas sector.

---

32. *The Petrobras Petroleum Engineering Educational System*, Otto Alcantara Santos.

Petrobras's work with CENAP has, through its programmes and further study courses, achieved some remarkable results. The company has not only increased the number of engineers in the local workforce, it has successfully created a safer and more efficient work environment. The company's own studies show that there is a direct correlation between the number of trained and certified people within the company and a drastic reduction in blowout incidents.

CONCLUSIONS

The collaboration and partnership between the S-system, the national education system and Petrobras has had a significant impact on educating nationals for the oil and gas industry. The numbers speak for themselves, and with enrolment figures high it is evident that Brazil's network for skills development is extensive, reaching even remote areas. Yet Brazil has also been open to engaging the international oil and gas workforce. This is evidenced by figures emerging from research conducted by BBC Brazil and the General Immigration Office of Brazil that reveal 49,801 foreign professionals entered Brazil during the period between 2010 and 2012 and were directly employed by the oil and gas sector. Overseas professionals are also employed by Brazil's oil sector to assist on research projects for the development of technology for pre-salt fields.[33] This openness to international participation has enabled Brazil to facilitate knowledge transfer within the workforce and has ensured that short-term skills gaps that may hinder industrial progress are addressed.

More broadly, Brazil has approached the challenge of skills and workforce development with an open mind and has borrowed wisely from the best of what the rest of the world has to offer. By aligning the activities of the national oil company with the mechanisms of state that are responsible for planning and developing skills for the labour market, the oil and gas sector has grown to become one of the most powerful on the planet. In the process, Petrobras has transitioned from being a national oil company to becoming an international operator with exploration and production interests all over the world. Petrobras has invested wisely in education, technology and research programmes that have allowed the company to exploit hydrocarbon reserves efficiently and have laid the foundations for a successful future. Central to this success, as previously mentioned, is the approach Petrobras has taken to investing in local people and improving the capacity of local education and training providers. The fact that IOCs have been known to 'poach' Petrobras-trained professionals due to their level of knowledge and

---

33. http://thebrazilbusiness.com/article/oil-industry-in-brazil.

technical expertise is a compliment to Petrobras' achievements in training and developing its people.

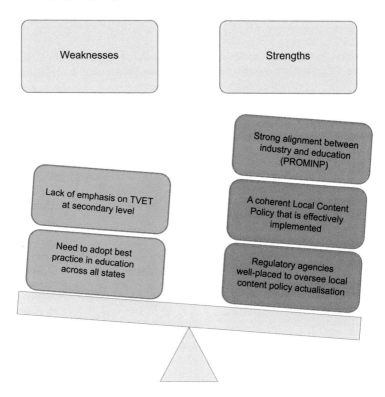

THE FUTURE

In 2009, when Brazil announced its new legislation to regulate pre-salt reserves, there was some doubt from the international community regarding the wisdom of changing how concessions were granted and around the establishment of a new state-owned company to oversee pre-salt oil and gas exploration. The argument put forward by critics of the changes and by Marilda Rosado, a former director of ANP and Petrobras, was that the previous system was transparent and fair, something that was important to maintain in Brazil, particularly at a time when corruption was becoming a major political issue. In 2014, the extent of Brazil's corruption problems was revealed and could have far-reaching implications for the wider economy. When police arrested Paulo Roberto Costa, Petrobras's former Chief of Refining, for money laundering, Brazil did not know that it would soon be watching a major scandal unfold. By February 2015, news reports showed that somewhere in the region of US$8.9 billion had been paid to politicians in bribes and kickbacks for awarding contracts. The scandal could damage the oil and gas sector in Brazil, reducing international confidence in the business climate. Petrobras is Brazil's largest investor. International credit agencies reacted to the scandal by relegating the debt of Brazil and its partially state-owned oil company to junk status. Brazil's economy cannot afford for the company to put a hold on spending. The future will depend on how policy-makers increase visibility and transparency. Petrobras has been an integral player in education and workforce development initiatives that have positively impacted the country at the grassroots level. It would be a shame to see that success reversed due to corruption in the higher echelons of the company and the political system.

Social unrest and protests – sometimes turning violent – have become a part of Brazilian life and are usually directed towards the government. This is emblematic of the fact that, while Brazil has made gains towards achieving social equality, there is still a tangible classism and a discernible chasm between the 'haves' and the 'have-nots', which is exasperated by the barely concealed nepotism rampant in the political system and extending into various sectors of the economy. Brazil will need to reflect on the Petrobras scandal and gauge not just the financial damage, but also what steps can be taken to prevent widespread political corruption. Brazil may need to take further steps to distance Petrobras from the federal government and make it less political. The Petrobras scandal has not undermined many years of positive work and progress in terms of workforce development and job creation. However, at a time of low oil prices, the scandal could not have hit the country at a worse moment. Although the government has announced plans to channel 100% of pre-salt revenues into education and health care, there is doubt as to whether the public will trust statements after the scandal.

Social unrest has not just been directed towards the federal government. Brazil is experiencing antipathy towards foreign business, particularly the IOCs operating in Brazil. IOCs must examine the reasons behind this and foster engagement with local communities. IOCs should seek to effect change and meet the expectations of people local to oil and gas production locations.

Brazil has a good pool of well-trained engineers, but not many of these are specialised for roles in the oil and gas industry. Engineering remains a popular route for many students. With expansions to the oil and gas sector expected, Brazil will need to consider how these students can be converted onto courses that have a petroleum focus.

Brazil's National Education Plan to increase spending on education to 10% of GDP by 2020 is commendable. However, analysts and experts have raised some concerns. Brazil's current spending on education is not low. However, the country will need to look closely at its education system and see how any future funds can be used more effectively and efficiently.

The Getenergy View

- Brazil has managed to build a national oil company into one of the largest energy companies in the world. Petrobras is one factor behind Brazil's economic success and has contributed towards it becoming a major world economy. At the core of this success is capacity development and the ability of the various mechanisms of state to work collaboratively in creating a system of education and training that is responsive and adaptable to the needs of industry.
- The oil and gas sector has benefitted hugely from the fact that from the 1940s onwards, the country developed a strong mechanism by which industry could interact with education in the form of the S-system. This approach is apposite for other countries looking to solve the challenge of connecting education with industry and offers a practical, replicable model of education and training provision for the oil and gas sector (and beyond).
- Petrobras's ability to adapt to accommodate new discoveries and technical developments in the sector is testament to the flexibility of its approach to education and training. Pre-salt discoveries have increased the complexity of the company's portfolio of assets but the company should be approaching these challenges with confidence.
- Petrobras has successfully adopted a culture of education that is encapsulated in its investments in and partnerships with universities and research centres. By having its own education and training facilities, Petrobras has ensured close ties to Brazilian tertiary education. In this way, Petrobras is able to align higher education with the needs of industry.
- Brazil's planned increases in education and training funding are commendable. However, there is little evidence to suggest that more money increases the effectiveness of education. The country needs to utilise its highly efficient education monitoring systems to ensure increased funding improves outcomes.
- The impact of education varies between the states of Brazil and efforts should be made towards decreasing this variance. There needs to be a greater awareness of where investments in education have paid off, and then a programme of implementing best practices nationwide. When that is achieved the planned increases in spending will amplify these successes. Without applying these lessons throughout the national education system, money will go to waste.
- To some degree public financing of education and training for the oil and gas sector is isolated from oil price shocks. This is for two reasons: firstly, education and national participation in the oil industry is politically driven. Secondly, Petrobras has a history of being resilient to oil price shocks due to – not in spite of – their dedication to education and training. If Petrobras wants to meet oil production targets for 2020 the company will need to invest in its personnel. History suggests that it will do exactly that.

Case Study 4

Iraq: The Struggle to Capitalise on Abundant Oil Wealth

Chapter Outline

This case study was made possible with the help of Josephine Clough and all the participants of the Getenergy Iraq Study Group.

INTRODUCTION

Iraq is the second largest producer of oil in OPEC and has the fifth largest proven oil reserves in the world.[1] Many of the country's reserves remain undeveloped, representing a huge opportunity for building the country's hydrocarbon sector and for creating economic growth and employment. Iraq has faced significant challenges in relation to the nationalisation of its oil and gas workforce. Against a backdrop of political and social instability as a result of successive wars and ongoing militant activity, the country has struggled to capitalise on its resource wealth.

Economic growth and future prosperity in Iraq is hugely dependent on the energy sector. This case study will explore the evolution of the country as an energy nation, analysing the efforts and achievements of government, the industry and the education system to develop and train Iraqis and to create a sustainable system of local participation. We do this within a context whereby the stability of Iraq as an energy nation will make a significant contribution towards global energy security and the ability of Iraq's nascent democracy to address the complex dynamics within the oil and gas sector – including those relating to skills and workforce development – will have a profound impact on the future of the country as a whole and on the well-being and prosperity of its citizens.

1. EIA, Iraq.

Education and Training for the Oil and Gas Industry: The Evolution of Four Energy Nations.
http://dx.doi.org/10.1016/B978-0-12-800974-1.00004-0

COUNTRY FACTS

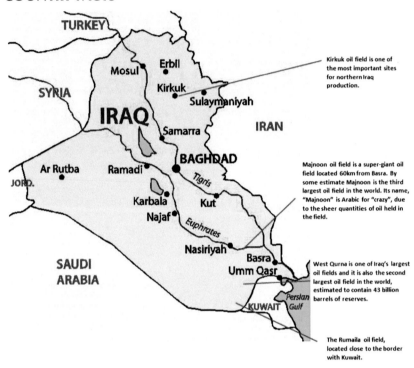

Kirkuk oil field is one of the most important sites for northern Iraq production.

Majnoon oil field is a super-giant oil field located 60km from Basra. By some estimate Majnoon is the third largest oil field in the world. Its name, "Majnoon" is Arabic for "crazy", due to the sheer quantities of oil held in the field.

West Qurna is one of Iraq's largest oil fields and it is also the second largest oil field in the world, estimated to contain 43 billion barrels of reserves.

The Rumaila oil field, located close to the border with Kuwait.

- In the south of Iraq there are five supergiant fields. Together these supergiants account for 60% of Iraq's total known oil reserves.[2] In the North, fields near Khanaqin and Mosul account for 17% of reserves.[3]
- Iraq's reserves are not equally distributed across the country. Most are located in the Shiite south or in the Kurdish north. Few fields are located in the Sunni west/central Iraq. This has been one factor behind sectarian conflict in the region, helping to foster Sunni resentment of the Shiite government and grow support for militant groups.
- In 2010, the oil industry contributed to 44% of GDP and 93% of total exports (other figures from the UN put this at 99%).[4] In 2014, the oil and gas sector accounted for 65% of the county's GDP – more than past years.
- Iraq managed to reach an output of 4 million barrels per day as of December 2014, but production has grown far slower than original government targets.
- The oil and gas sector employs 1% of the total labour force.[5]

2. International Energy Agency, *World Energy Outlook Special Report: Iraq Energy Outlook*, (October 2012).
3. Ibid.
4. Iraq, Ministry of Planning, 2010.
5. UNDP.

- In 2013, the Parliament's Economy and Investment Committee announced an unemployment rate of 25%. The true figure is likely to be much higher than this.[6]
- Young people (15–24) are the most affected by unemployment in Iraq (18%[7]) and with 450,000 young people joining the workforce annually this figure is set to worsen.[8]
- Among young people, females are the most affected by unemployment and, more generally, young people entering the workforce from higher education are less likely to find employment.
- 24% of the workforce is uneducated – 43% of people only have primary school level education. These figures show the extent to which skills development is becoming a major concern for the country.

HISTORY OF OIL AND GAS INDUSTRY

Oil wells of the Iraq Petroleum Company in the Kirkuk District in northern Iraq in 1932.
(© Shutterstock, Everett Historical.)

The quest for oil in the region now known as Iraq has its beginnings in the Ottoman period. In the late 1800s, European corporate entities believed that Mesopotamia (as Iraq was then known) could hold considerable oil reserves. Deutsche Bank, Royal Dutch Shell, the National Bank of Turkey and the British Government-controlled Anglo-Persian Oil Company came together to create the Turkish Petroleum Company, the forerunner to what would later become as the Iraq Petroleum Company (IPC).

Despite the Turkish Petroleum Company being formed in 1912, it would not be until after World War I and the carving up of the Ottoman Empire that

6. Iraq, Workforce Development, *SABER Country Report 2013*.
7. Official UN figures.
8. Education and Labour Market Programme for Iraq, *Action Fiche for Education and labour Market Programme, 2013*.

the company would go on to explore and find oil in Iraq. In 1925, the Turkish Petroleum Company was granted rights to explore for oil with the caveat that the Iraqi government would receive a royalty for every tonne of oil extracted. These royalties were linked to the earnings of the oil companies and were not payable for the first 20 years. Oil was first discovered and successfully extracted in 1927 at a well located in Baba Gurgur, a large oil deposit near Kirkuk in Northern Iraq.

By 1928, the shareholders of the Turkish Petroleum Company signed a partnership agreement to include the Near East Development Corporation (NEDC), which was an American consortium of five large oil companies (Standard Oil, which would later become ExxonMobil, Standard Oil Company of New York, which too would be absorbed into the ExxonMobil amalgamation, Gulf Oil, later to merge with Chevron, the Pan-American Petroleum and Transport Company and Atlantic Richfield Co., which would join BP in 1999). Effectively this coalition of companies was a cartelisation of Middle Eastern oil in ex-Ottoman territories. The agreement came to be known as the Red Line Agreement and it represented a binding contract between all partners and prohibited shareholders from operating independently of the Turkish Petroleum Company in strategic Middle East regions. In 1929, the Turkish Petroleum Company was renamed the Iraq Petroleum Company (IPC) and the Red Line Agreement gave the IPC power to monopolise operations in the region.

Since the collapse of the Ottoman Empire, Iraq's oil has been a cause of political tension. During the early 1900s, the Turkish Petroleum Company enjoyed good government relations with the Hashemite monarchy under King Faisal I. While the Hashemite monarchy did not challenge over oil rights, concessions and agreements, they did raise other issues: the need to increase production (which during the Great Depression was minimised so as to prevent the oil glut worsening and Iraqi oil forcing the price even lower) and to increase local participation in extraction operations. The Turkish Petroleum Company's strong influence over the monarchy meant that production could be increased or decreased according to international demand and market prices. Further to this, local participation in the oil industry was kept to a minimum to avoid nationalisation, allowing the Turkish Petroleum Company to maintain its position and influence in Iraq. Consequently Iraq's petroleum industry was one that reflected international business needs and not the national agenda.

The tight control of Iraq's hydrocarbons by foreign corporate entities and international oil companies did not go unnoticed by the Iraqi people who, by 1958, rebelled against the established monarchy and took power through revolution. The IPC found it difficult to maintain its tight control in the face of these revolutionary governments who saw how vital hydrocarbons were to the Iraqi economy. Attempts to wrest some degree of control over Iraq's oil and gas industry by Abd al-Karim Qasim, a nationalist Iraqi Army general who took power through a coup in 1958, were thwarted by the IPC. Knowing that Iraq did not have the technical or managerial expertise to assume direct control over oil production operations, he instead tried to raise transit rates in Basra – an important shipping point for oil in the region. The IPC responded to this by suspending production in areas that relied on Basra as a transport hub.

A major turning point came in December 1961 when the Iraqi government passed Law No. 80, allowing the government to take control of 99.5% of IPC's concessions. This meant that the oil companies who were partners in the IPC were limited to their existing facilities and production areas, or just 0.5% of their original concession. Most importantly, these concession areas were taken without compensation and led to an immediate cessation of oil exploration activities. This time represented a shift in government relations with IOCs seeking to profit from Iraq's oil and gas reserves and the control they had until then exercised over extraction operations, concession areas, intermediate fees and industrial development.

After a second coup that ended in the death of Qasim in 1963 and the establishment of the Ba'ath regime, the new Ba'athist government refused to negotiate Law No. 80, stating that its implementation was final. In 1964, the Iraq National Oil Company (INOC) was established to help facilitate exploration and production activities and foster local participation in the oil and gas industry. Agreements were arranged and deals struck between the IPC and the government. Under the presidency of Abdul Salam Arif – who was a key protagonist in the overthrowing of Qasim's government in 1963 – the IPC was granted an additional 0.5% concession covering the North Rumaila field – an extensive field that had remained undeveloped during IPC control of Iraq's hydrocarbon assets. After some debate, it was decided that the IPC should not have control over the North Rumaila field and in 1967, Law 97 was enacted giving the Iraqi NOC the exclusive right to develop oil in the country, forcing IPC to suspend operations. A month later the government issued Law 123, consolidating INOC's power and exclusive rights over the oil sector.

In 1968, an agreement was struck for oil fields to be developed in conjunction with the Soviet Union in a deal known as the Iraq-Soviet Protocol. A number of other actions were taken to help Iraq continue oil and gas production including the restructuring of INOC. INOC would be overseen by the Ministry of Oil and would receive a mandate to enhance, expand and control the production of hydrocarbons at every stage of the process from upstream to downstream. Lacking the technical and managerial experience required, INOC and the Ministry of Oil turned to France and the Soviet Union for help in implementing its oil policy. By the 1970s, the Iraqi government nationalised all of the IPC's assets thereby completing the nationalisation of the oil industry. The nationalisation process had put the country at odds with international operators and led to political tensions with Western governments.

European and American oil companies, operating through the IPC, brought about the conditions for aggressive nationalisation. In manipulating output, keeping production low and forcing the government to be dependent on royalties from concession areas, the IPC kept the citizens of Iraq from ever seeing the benefits of its oil. This assured political upheaval and – as has been demonstrated in the preceding paragraphs – the constant coups and revolutions that eventually led to the IPC losing its grip on Iraq's hydrocarbon-producing facilities. While Iraq's actions were not universally judged to be fair and just, it must be asserted that this was a pivotal point for the oil and gas industry in the Middle East and had significant ramifications for relations between IOCs and governments seeking to expand their economies through hydrocarbon extraction and production.

Unfortunately, under Ba'athist rule very little was done to develop the oil industry. Attempts at workforce development are best understood by looking at exploration and production activities during this period: Iraq only drilled 2000 oil wells under the Ba'athists.[9] In comparison, Texas drilled over a million wells during the same period. Under the regime of Saddam Hussein, the fifth President of Iraq, the government controlled over 80% of the economy, making 1.2 million people wholly dependent on government employment and a further 300,000 dependent on government-paid pensions. Two in three households were directly dependent on the government for income.[10] This meant that the oil industry did not play a significant role in achieving the goal of economic independence directly sought by successive governments who had pushed for nationalisation. Iraq's oil sector became neglected and suffered under authoritarianism. This left oil production facilities and refineries in a state of chronic disrepair. Previous governments with access to oil revenues had ignored the need to develop human resources and failed to invest in education and training initiatives. There are suggestions that previous governments used oil funds to support 'political patronage'.[11] It is also argued that having resources in the ground can make governments less likely to tap into the resources of its people. Rentier states – often called because they rent access to natural resources to international corporate entities – do not tend to utilise local human resources. For Iraq, the nationalisation of the oil industry did not achieve a higher level of local participation, nor did it mean that larger investments were made into developing the technical and vocational skills required to operate and manage future extraction and production operations.

Without clear and realistic development policies, the Ba'athists did little to address the systemic issues that originally led to the nationalisation of the oil and gas industry. Despite the government's reliance on oil revenues, it continued to neglect aging pipelines and production facilities and failed to invest in technical

---

9. Oil and Democracy in Iraq, *SOAS Middle East Series*, pp. 35.
10. Ibid
11. Giacomo Luciani and Felix Neugart, "Toward a European Strategy for Iraq", *EUI Policy Paper* (March 2003): 6–7.

or managerial skills. Under Saddam Hussein's regime, Iraq once again became dependent on IOCs. Companies such as TotalFinaElf (renamed Total in 2003) managed to negotiate significant contracts for rights over production in the Majnoon and Nahr Umar oil fields in the southern region of Iraq. In 1997, the Russian company Lukoil agreed on a 23 year contract on the West Qurna field, upgrading facilities to increase production. Additionally Russia's Gazprom was involved in repairing gas stations in the early 2000s. The China National Oil Company agreed to a deal allowing the company rights to explore oil in the Al Ahdab field over a 22-year period.[12]

By the 1990s, under Saddam Hussein's presidency, PSAs (production sharing agreements) were granted to Russia, China and a number of other international players in the oil industry. These PSAs provided Iraq a profit percentage of under 10%.[13] In 2007, a new law (the 'Iraq Oil Law') was drafted. This law proposed to give foreign companies a favourable legal framework to begin operations. The New York Times commented on the controversial law, noting how it promised to change the landscape of Iraqi oil production, making INOC impotent and giving it just 17 of the then 80 known fields. Additionally IOCs would not be required to invest in the local economy and much of the revenue would be pocketed by the companies, flowing away from Iraq:

> *The foreign companies would not have to invest their earnings in the Iraqi economy, partner with Iraqi companies, hire Iraqi workers or share new technologies. They could even ride out Iraq's current "instability" by signing contracts now, while the Iraqi government is at its weakest, and then wait at least two years before even setting foot in the country. The vast majority of Iraq's oil would then be left underground for at least two years rather than being used for the country's economic development.[14]*

International operators, under this legislation, were given an 'easy ride', and were only required to meet basic corporate social responsibilities. There was also no requirement to invest in the development of local people for the oil and gas sector. In 2006, under former Prime Minister of Iraq (2006–2014) Nouri al-Maliki, the Ministry of Oil considered awarding no-bid contracts to ExxonMobil, Shell, Total, Chevron and BP. In 2008, this was scrapped after labour unions sent a letter to the Iraqi President stating that 'production-sharing agreements are a relic' and 'they will re-imprison the Iraqi economy and impinge on Iraq's sovereignty since they only preserve the interests of

---

12. Trish Saywell, "Oil: The Danger of Deals with Iraq", *Far Eastern Economic Review* (March 6, 2003).

13. Janabi and Ahmed, "Row over Iraq Oil Law", *Al Jazeera* (May 05, 2007).

14. New York Times, Whose Oil Is It Anyway? (March 13, 2007). http://www.nytimes.com/2007/03/13/opinion/13juhasz.html?_r=0.

foreign companies'.[15] Under ex-Oil Minister Hussein Shahristani Iraq held licensing rounds in 2009. Companies were invited to bid on technical service contracts (TSCs) where companies were paid based on a per-barrel fee of extracted oil. This has created a situation where international companies operating in Iraq today are essentially contractors to the government with no stake in exploration and production activities. This diminishes the likelihood that these companies will see the benefits of investing in long-term education and training projects or contribute to local capacity building. Iraq has moved between two extremes when dealing with oil companies, neither of which have proven to be in the best interests of the country.

In considering Iraq's oil and gas history, we see that

- Iraq is an important country for the international oil industry. Since World War I, IOCs have sought to participate in the production of oil in Iraq and this has resulted in IOCs becoming embroiled in the politics of the country, creating a fractured relationship between local communities and the industry.
- Iraq has historically tried to alleviate its dependency on IOCs through the enactment of laws and policies that limit the control of operators. This led to the establishment of INOC, which eventually would be charged with driving forward Iraq's oil and gas industry and furthering the country's interests in this area. However, their efforts were hampered by a lack of skilled personnel and poor leadership.
- Laws 80, 97 and 123 were especially important to this story and represent pivotal points in Iraq's hydrocarbon history. These laws enforced sovereign rights over oil and gas. They gave more power to the Iraqi people and removed the IOCs from their position of control over the country's oil and gas sector.
- Despite these laws and policies, Iraq would become a rentier state, renting access to oil reserves to operators at very little benefit to itself. Oil revenues would be used to support the ruling government instead of being used for investment in initiatives to broaden the economy and support growth and workforce development.
- Commentators and analysts argue that Saddam Hussein and the Ba'athist party were responsible for Iraq's rentier state economy. It is further argued that oil price increases in 1973 and 1974 helped strengthen the ruling Ba'athists. Generally when a country becomes a rentier state and dependent on natural resources it removes responsibility from the government to develop skills and human resources.
- After the 2003 war, the government would draft new laws that would again engage international operators in Iraq's oil industry. After protest from intellectuals, political figures and the Iraqi people, no-bid contracts were scrapped and IOCs were invited to bid on service contracts where payment was based on a price-per-barrel agreement. After some initial problems with prices, leading to frustrations among IOCs, deals were struck and IOCs again became active in the region.

---

15. IPS, *Iraq: New Oil Law Seen as Cover for Privatisation* (February 2007). http://www.ipsnews.net/2007/02/iraq-new-oil-law-seen-as-cover-for-privatisation/.

Dropping a welded pipe into a trench with a crane while laying the Iraq Petroleum Company's pipe line across Palestine's Plain of Esdraelon, July 1933. Completed in 1934. (© *Shutterstock, Everett Historical.*)

HISTORY OF EDUCATION, TRAINING AND WORKFORCE DEVELOPMENT

Prior to the Gulf War (1980–1988) and the Persian Gulf War (1990–1991) Iraq enjoyed one of the strongest systems of higher education in the Middle East. In fact, the importance of education is enshrined in Iraq's ancient history, where Mesopotamia – the ancient name for Iraq – was the seat to several important civilisations and the home to some of the world's great rulers.

During the Ottoman period, the regions of Basra, Baghdad and Mosul, which today make up modern Iraq, were largely untouched by the cultural and educational revival that swept through Constantinople (officially renamed Istanbul by Ataturk in the 1920s) and other parts of the empire. Education at the collapse of the Ottoman Empire was directed by Mullahs in small schools known as Kuttab, which served to provide young children with a very basic education alongside strong religious fundamentals. Abd al-Karim Qasim (Prime Minister of Iraq 1958–1963) was concerned with uniting the Arabs and Kurds. Education was seen as an important means of bringing ethnic and minority groups together and unifying Iraq's many people and forging a single national identity. It was during the early Ba'athist regime - where the concept of "Arab Socialism" became entrenched - that Iraq saw the greatest progress in terms of education reform.

Iraq's nationalisation of the oil industry could not have been better timed. Soon after complete nationalisation was achieved the oil boom of 1973–1974 took hold and helped Iraq flourish, achieving economic growth and facilitating social change. The key turning points relate to the establishment of three highly important laws: an Illiteracy Eradication Law (1971), a Free Education Law (1974) and a Compulsory Education Law (1976). These new laws had a cumulative effect, driving up education standards. The laws guaranteed all Iraqis received an education and committed the state to providing free meals to school children, as well as adequate education resources and facilities. In 1978, the government took further action – the National Comprehensive Campaign for Compulsory Literacy Law was enacted and specifically addressed the teaching of reading, writing and arithmetic and the development of professional skills.

In order to facilitate the huge education initiative that swept across Iraq – even in the most rural areas – women were encouraged to participate in the education system, becoming teachers, joining universities and undergoing vocational training. Many academics have acknowledged the Ba'athist regime's success in helping women join the workforce and in guaranteeing the general freedom of women. At universities they were able to mix freely with men and learn and teach at the same institutions. In the 1980s, women accounted for almost 50% of teachers.[16] It was during the early years of Saddam Hussein's rule that these and other goals were achieved for Iraq's education system. However the subsequent wars would reverse much of this.

Up until the end of the 1980s and beginning of the 1990s, Iraq spent more on education than most other developing nations (up to 5% of the national budget). Students from Africa and across the Middle East saw Iraq as a preferred place to study. This reputation was steadily eroded by wars and sanctions. Primary attendance was compulsory until 2000. During the period between the beginning of the war and the withdrawal of US troops in 2014, estimates indicate the deaths of skilled teachers and lectures ran into the hundreds due to ongoing security issues and the rise of militancy in the region. 80% of assassination attempts were on university staff, with lecturers and assistant lecturers particularly at risk.[17] Universities gained a reputation for being dangerous. Security concerns had two primary effects: firstly, students did not want to continue enrolling on courses in places where they could be targeted in a terrorist attack; secondly, instructors and teachers were concerned that they were employed by institutions at risk of attack (to illustrate this threat it was found that a substantial number of schools were captured by *Islamic State* militants in 2014[18]). Additionally many students were forced out of formal education to support their families who could no longer afford the expense of educating their children. Issues with roads, buildings and other basic infrastructure created further barriers to

---

16. Amal al-Sharki, "*The Emancipation of Iraqi Women*", in *Niblock* (Iraq), 80–81.

17. Mohammad al-Hourani, "Iraqi Academics: Between the American Hammer and the Israeli Anvil" (in Arabic), *al-Fikr al-Siyasi (Political thought)* 8, no. 24 (2006).

18. Information from Getenergy VTEC MENA 2014, Iraq Study Group – www.getenergy-intel. com

those wanting to continue their education and a lack of financing continues to plague the system. This is particularly true of the technical and vocational education system.

We can see from Iraq's education history that

- Education in Iraq was, at one time, viewed as the best in the Middle East region. Iraq enjoyed high attendance at all levels of the education system and managed to attract students from across the Middle East and North Africa. Up until 2003, attendance was compulsory and students were able to progress through the system.
- There was and remains a lack of effective Technical and Vocational Education and Training (TVET) provision in Iraq. More specifically, training for the oil and gas industry is sparse and there are few options available to Iraqis seeking to pursue a career in the oil and gas sector.
- Without the due attention to developing an effective TVET strategy within a wider education policy Iraq will continue to lack the oil and gas professionals (engineers and technicians) required to develop a strong economy.
- Wars and unrest have contributed significantly to the erosion of the education system at all levels in Iraq, displacing students and teachers and creating physical barriers to attendance.
- These wars have made institutions of education unsafe areas. Universities and schools are often the target for militants and come under terrorist attacks. Anecdotally even the technical schools opened by the IOCs seeking to increase local participation are sometimes threatened by attacks. In a country where people are afraid to attend classes and enrol on courses at institutions, little progress on skills development can be made.
- TVET in Iraq suffers from a lack of skilled teachers and instructors who know and understand the industry for which they are teaching and training.

It is against this backdrop that we consider the efforts of government, industry and others to develop a skilled workforce for Iraq's hydrocarbon sector.

WORKFORCE AND SKILLS DEVELOPMENT FOR THE OIL AND GAS INDUSTRY

For many years, the oil and gas sector in Iraq was effectively under international control. During this time, there was little investment in the education and training of local people with international companies pursuing a familiar pattern of employing expatriates from other parts of the world to undertake the technically challenging activities demanded by their exploration and production projects. During the transition to oil nationalism, the country benefitted from the help and guidance of the Soviet Union, particularly in relation to the technical aspects of exploration and production. And whilst this period undoubtedly saw the rise of the Iraqi oil and gas professional (with those working for the national oil company effectively learning on the job) there is little evidence of any long-term skills development strategy to go alongside efforts to nationalise the industry. The failure to address longer-term issues of workforce development was crystallised when the government of Sadaam Hussein felt compelled to invite international operators back into the country in order to bolster what had become an underperforming industry. The historical neglect of skills and workforce development in Iraq now presents a systemic problem to the oil and gas industry. Comparatively few Iraqis work in the oil and gas industry today and there is little by way of dedicated petroleum programmes for locals to enrol on.[19] The fragmented nature of the oil and gas sector coupled with a failure to create a coherent national skills approach has left Iraq with a deficit of skilled technicians and has created an overreliance on expatriate workers. In understanding why this situation has arisen we can look at the approach taken to technical and vocational education and at how this approach has failed Iraq and its people.

In an economy fuelled by the oil and gas sector, local participation should be a primary aim for governments and citizens alike. The industry offers employment at every level – from technicians and engineers to leaders and managers. The history of the oil and gas sector in Iraq shows us that the country has a complex relationship with IOCs but that underpinning this is the reality that without these companies – and their technical and managerial expertise – the industry in Iraq struggles. One of the key reasons for this struggle lies in the fact that Iraq lacks a comprehensive education and workforce development strategy and a clear means through which education can align itself to the industry. There is very little in the history of the industry – and the concurrent history of education and training in Iraq – that suggests a coherence between industrial growth and educational strategy. Furthermore, the historic strength of the higher education sector in Iraq may have diverted attention away from the need to invest in skills development. Educational pathways into vocational and technical education have not always been clear (and bright students have not been encouraged to pursue these pathways). Furthermore, the country has failed to develop the science and mathematics skills that provide students with the opportunity to progress through the education system and into university and that are the bedrock for employability in the industry.

Despite challenges – and despite the ongoing barriers to Iraqis taking up positions in the sector – experts working in-country acknowledge that there is real demand amongst locals to work in the industry and particular interest in engineering programmes. International operators have little trouble sourcing local people from colleges and universities because of their willingness to pursue oil and gas careers. However, there are still large numbers of expatriates working in the industry throughout Iraq indicating that the challenge here is not one of quantity but quality.

Analysis of Iraq's recent past in regard of workforce development perhaps offers us the most productive lens through which we can peer. Activities over

---

19. UNDP Iraq, *UNDP Iraq – Shell Partnership, UNDP Private Sector Focal Points Meeting 2013* (Geneva, April 2013).

the last 10 years - and plans for the next 10 years - tell us much about the challenges Iraq has been struggling with and about the strategy it hopes will address these challenges. We might begin this exploration by looking at the launch of the government's National Development Policy (NDP) 2010-2014 which aimed to achieve stability 'under the auspices of a federal democracy' and 'in accordance with market mechanisms'.[20] The NDP further stated that human development should be based on assessing market needs and creating employment for individuals. Much of this hinges on job growth and improving the quality of the workforce. Education standards are a critical part of the NDP and related documents note the importance of connecting industry demand with education and training. In particular, the lack of skilled Iraqis for the oil sector presents a real challenge. Royal Dutch Shell published an internal analysis of the workforce needs in Iraq, which revealed that 5000 skilled workers would be needed between 2012 and 2017. These 5000 workers would be required for jobs in technically intensive areas such as welding, pipefitting and rigging. The report went on to suggest that skilled professionals would need to be educated vocationally and that institutional capacity had to be improved in order to meet the need for instructional expertise and facilities. This demonstrates both the demand for skills that exists and, simultaneously, the challenges of meeting this demand locally.

Alongside the NDP, the government also produced a National Strategy for Education and Higher Education in Iraq, 2011–2020. One of the objectives for the strategy is to increase enrolment in TVET from 2% to 10% in the run up to 2020. At present there are a number of barriers preventing progress in this regard. With specific reference to the education and TVET system these challenges include the following:

- There are issues of overlapping responsibilities between ministries. This especially hinders education for the oil and gas industry because the Ministry of Oil holds responsibility for petroleum education while the wider responsibility for education is divided between three other ministries.
- There is weak organisation among decision-makers due to a highly centralised administrative structures within government and relevant ministries.
- There is a culture of slow decision-making on the part of the various authorities responsible for the education sector meaning that the system is not responsive to the needs of industry.
- The complexity of laws governing education and skills creates restrictions for those engaged in trying to improve provision.
- Teachers and students have been displaced by wars and upheaval creating a situation where many learners cannot access formal schooling due to teachers and students being physically unable to reach their institution.
- Shortages of resources and supplies, coupled with neglected infrastructure (and student overcrowding due to lack of space), have placed significant pressures on the education system.

---

20. Ministry of Planning, National Development Plan 2013–2017.

- There is an acute lack of teacher performance evaluations and quality control measures across the system.
- TVET is missing a clear plan of student progression – with students entering vocational courses unable to access further educational pathways thereby diminishing the attractiveness of assuming a role within a vocational institution.

Specific analysis of TVET provision in Iraq reveals that there needs to be a greater emphasis on the development of TVET as a strategic subsector of education in order to better prepare the workforce for technical positions. This is especially important when taken alongside international and national hopes for Iraq's oil industry. The dynamic between the TVET sector and the oil sector is absolutely critical in this regard – if the country can achieve its objectives here, then other sectors of the economy can follow. However, the TVET system is struggling to overcome serious, systemic issues that include:

- Low rates of enrolment due to the unpopularity of vocational education as against academic study
- Outdated, didactic teaching methods
- Poor quality training facilities and equipment that is not fit for purpose
- Teachers who lack the pedagogical ability, skill sets and required industry experience to deliver an adequate learning experience
- A lack of alignment between education and employers
- A TVET system that does not have the needed industry guidance to formulate structures of quality control and frameworks for competencies and qualifications
- An unstable market that lacks clear objectives and routes for the provision of training

Further studies of Iraq's TVET system indicate that the country needs a 'supply-driven system' that seeks to meet the needs of the labour market and, overall, combat the debilitating problems that cripple Iraq's economy. This objective sits within a wider set of strategic goals for Iraq that include the elimination of poverty, regional stability, reduction of unemployment and more opportunities for women to enter the workforce.

The World Bank's 2013 report assessed Iraq's effectiveness in terms of skills and workforce development. The Systems Approach for Better Education Results (SABER) looks at Iraq from three dimensions:

- Strategic framework
- System oversight
- Service delivery

Iraq was rated at the latent stage for all three dimensions, the lowest category on the SABER analysis. The report noted that a national qualifications framework (NQF) was not in place and that without a qualifications framework there could be little harmony between the oil and gas industry, or indeed any industry, and the education system. The report also stated:

> *[in] absence of identified economic prospects and their implications for skills, there is no clear strategy or plan of action for WfD [workforce development]. The lack of an executive commission to coordinate actions of WfD stakeholders, effective and efficient funding mechanisms, and quality control of training provision also explain the current level of development in Iraq's WfD system.[21]*

Further to this there are no 'clear agendas, working protocols, legal commitments or consistency of work' amongst authorities in Iraq. This report – alongside the other analysis quoted here – reveals the depth of the challenges that face Iraq today and also the historic failures to establish the mechanisms of government to create coherence between industry, education and the labour market. It is within this context that a failure to nationalise the oil and gas workforce exists.

There have been further efforts in recent years to tackle the endemic failures of the skills development system. In 2010, a TVET High Committee was established in the Advisors Commission within the Ministerial Council. Made up of a number of TVET stakeholders – including the MoE, MOHESR, MOL, MOP, Iraqi Businessmen Union (IBU) and Iraqi Federation of Industries (IFI) – the council involved many of the right people required to create change within the Iraqi TVET system. However, this group had no legal status and no decision-making power, significantly limiting its ability to overhaul the TVET system or to drive forward workforce development. The World Bank has suggested that this committee be given 'clear mandates and responsibilities regarding system oversight'.[22] Iraq's authorities must address the problem of creating effective training and education policies to facilitate a demand-driven approach to technical and vocational training. The National Strategy for Education and Higher Education in Iraq, 2011–2020, makes clear suggestions and recommendations for how this might happen (although the recommendations are broad and it is not clear how they will be achieved). It suggests that:

- improving the public perception of TVET is critical
- any attempts to train and educate a workforce for the oil and gas industry will be moot without a clear national qualifications framework (NQF)
- education and training must be standardised across the country
- ensuring that private and public training institutions have the needed technical equipment is critical. Information systems for engineering and geoscience courses, simulators for rig work and safety training and other important equipment are all essential to the quality of education.

---

21. Iraq – Workforce Development, World Bank, SABER Country Report, 2013.
22. Ibid.

- introducing innovative measures to educate the oil and gas workforce, including online and distance learning to combat the low levels of attendance at universities and institutes, may move things forward.

There is now a law in place to establish the Federal Service Council. This Council will help individuals to find jobs in the private sector and assist with career development. However, action on establishing this Council has been slow, fuelling the sense that a failure to effectively implement policy is a significant stumbling block to progress.

Analysts have also noted an additional challenge to the effectiveness of current policy initiatives. Iraq has not enacted legislation that clearly defines the responsibilities of IOCs in developing local content. Strictly speaking, oil companies are not legally accountable for how they engage the local population in extraction and production. Ideally the government should be making significant steps towards an extensive hydrocarbon law that lays out the roles and responsibilities and there must be a coherent local content policy in place to which oil companies are held accountable. Draft legislation does exist, but concrete steps must be taken and laws must be passed as a matter of urgency. The fact that IOCs do not have an equity stake in oil production within Iraq may also impact on their willingness to lead training initiatives (see 'structure of the industry' diagram below).

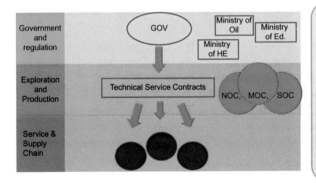

Structure of the Industry.

There is evidence that, despite the lack of incentives offered, some IOCs are willing and ready to train locals and educate them for the industry. Lukoil Overseas has established a training centre in Basra province. This centre is the result of collaboration with Basra Technical Institute. The two partners are working to improve core skills and knowledge in science and mathematics, a critical factor missing since 2003, leaving many Iraqis without the basic education they need for employment. Other operators, such as Shell, are also working to improve the local skills pool in a similar way. But these attempts are largely in isolation – they typically serve the companies themselves and do not exist within a wider framework. This means that the capacity often resides within the companies themselves and so the skills and infrastructure developed will disappear if companies choose to move on. As a consequence these initiatives have little impact on combatting wider systemic issues of workforce development. Capacity building requires a holistic partnership between the government and the IOCs and currently the work undertaken by IOCs is self-defined and self-motivated. It is incumbent on the government to create the framework within which these individual initiatives can deliver collective momentum.

The fractured nature of efforts towards workforce development calls for more systemic thinking. Failure to embrace long-term investment alongside a transparent means of investing oil revenues has hampered any meaningful progress on education reform. Earmarking profits for infrastructure investment would force the various ministries responsible for technical and vocational training and training for the oil and gas industry into collaboration. There has also been a collective failure to provide clear legislation on local content and to implement within the education and training sector a coherent qualifications framework that adheres to international standards. If this is accompanied by support for a strong national oil company and a local content policy that offers clear guidance on the responsibilities of international operators then impact can be achieved. An Iraq where national interests are protected but international competition compels the free market economy to strive towards new heights should be the vision of Iraqis, Kurds and the international community.

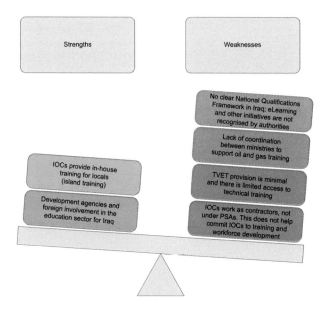

In conclusion the following points summarise the views of those currently engaged in workforce development for the oil and gas industry in Iraq:

- IOCs are enthusiastic about hiring local Iraqis for operations in the country. This is not only down to a sense of social responsibility - there are cost benefits to hiring locals and IOCs want to take advantage of this.
- A major concern for IOCs is a mismatch between industry expectations and what educational institutions provide in terms of training. IOCs are finding that they still need to give college graduates training in the basics. There is a lack of relevant experience and expertise amongst graduates and this is a reflection of the curriculum and the facilities, both of which require improvements.
- IOCs are finding it difficult to execute plans for a joint-industry training centre that could help to improve the access to, and the quality of, TVET in the county. While IOCs are eager to begin implementing their plans the government has been slow to respond to the growing capability gap.
- IOCs have a bleak view of the future of Iraq's TVET system if more is not done to improve local capacity building.
- Power must devolve from the government to a larger group of individuals responsible for curriculum development and decision-making for TVET provision. Too often, senior government figures change their views as to what the aims are and how to achieve those aims. This leads to a misalignment between industry and the education system.
- There are considerable problems around language ability. IOCs cannot replace oversees staff with locals until they are able to understand English.
- IOCs are certain that they share a common interest in capacity development with the government. Yet there is little alignment to address skills shortages, curriculum challenges and the provision of education. A fundamental question must be asked: if there is a common interest, why is it proving so difficult to align industry with government?
- IOCs do not have an equity stake in oil production. The Iraqi government prefers to use Technical Service Contracts (TSCs) instead of Production Sharing Agreements (PSAs). This is partly due to a history of bad relations with IOCs, who for many years manipulated the oil industry to further their international business objectives. It is argued that TSCs do not encourage investment from IOCs and consequently makes them hesitant to lead any training initiatives. This may have to be addressed if the challenges of workforce development are to be tackled.
- A forum for engagement does not exist – there are no structures through which IOCs, government and education can engage in meaningful dialogue and there is little horizontal or vertical alignment between education institutes and the ministries overseeing them. Neither the oil industry nor the government is proactive in creating engagement through which actions can be agreed, information shared and development objectives achieved.

A river of crude oil flowing through the Iraq desert resulting from a gusher Kirkuk District of Iraq.
(© Shutterstock, Everett Historical.)

IRAQ'S FUTURE

Iraq's challenges are numerous, but hopes are pinned – as they have been since the fall of Saddam Hussein's regime – on the oil industry. With huge potential to sharply increase output and consequently increase revenue into the country the question is now one of workforce development and improving local capacity. How can this be achieved effectively, what role do IOCs play and how can they work in collaboration with an impotent NOC and fragmented government to jointly address the issue of workforce development? Further to this, the challenge ahead will largely depend on how Iraq avoids authoritarianism and how the local population takes ownership of the oil sector. The answer to these questions and challenges is important for two reasons: firstly the oil sector is of national importance and has the potential to lift the country out of the crippling circumstances that have hindered social and economic development; secondly Iraq is not only important to its immediate neighbours but also to the entire global energy system, where increased production is necessary to meet energy demand. As fast-growing economies, such as China and India, continue to expand, energy demand increases. It has been estimated that every percentage point of economic growth brings with it a 0.5% increase in electricity demand.[23] Iraq's future as an energy nation will depend on the following:

- Local content must now be addressed. Development agencies and political bodies all acknowledge that Iraq has not given this important area due consideration. The private sector can assist workforce development, but this can only be achieved if economic liberalisation reforms are made. If the private

23. Keeping the lights on, Mark Johnson, "The World in 2015", *The Economist* (December 2014).

sector - and especially IOCs - play a prominent role in shaping the design of education, this will ensure greater levels of employment leading to real economic growth that is not dependent on the extraction and exportation of hydrocarbons. Part of this will involve a strong local content policy married to a national development plan that sets realistic long-term goals. A national development policy now exists and steps are being taken towards implementation of the policy. Alongside education the government must consider how knowledge transfer and technology development will be addressed as part of the approach. Systems of accountability must be put in place to monitor adherence to local content policy by international operators. There should be an independent regulatory entity established to monitor, report and address local content.

- Internal problems concerning oil ownership between the Kurdistan Regional Government (see additional information at the end of this case study) and the Baghdad government threaten to escalate conflict in Iraq. Both sides lay claim to Kirkuk, an important oil-producing area that both the Kurds and Iraqis want to control. The threat of Islamic State has brought this region into focus as the Baghdad government and Kurdistan Regional Government seek to bring the oil-rich area under their respective spheres of influence. The Islamic State provides a useful pretext for Shiite militias, loyal to Baghdad, to enter Kirkuk and possibly reshape the demographics of the area through the redistribution of property, handing buildings and housing to Arabs who will support the government in Baghdad. Possible situations like this could lead to clashes between Kurdish Peshmerga forces and Shiite militias. The future of oil production will largely depend on regional stability, and a compromise on energy exports being reached between Erbil and Baghdad.

- Iraq will need to earmark future oil funds to ensure sustainable investment in education and training, especially at secondary and tertiary level, is bolstered through government initiatives supported by – and perhaps even led by – the oil industry. IOC involvement cannot be limited to extraction operations but must, ultimately, involve plans to work with the relevant ministries to support the development of the education system. Some think tanks favour a private decentralised model for Iraq's oil revenue, where funds are handed directly over to citizens. This approach has worked in Canada and it can help to directly impact on households and individual citizens. But there are questions pertaining to how this might work in Iraq and whether this option would benefit the country as a whole. Citizens need basic utilities and social necessities, not individual wealth. Oil funds must be managed in a way that is transparent and meets the requirements of international legislation. These funds should be channelled towards social development and infrastructure upgrades.

- It is hoped by many – especially international operators in Iraq – that a joint, industry-led training facility can be established. This centre will be a joint venture between operators, but before training and development initiatives such as this can be implemented fragmentation amongst the various ministries of the Iraq government must be addressed. IOCs are willing to cooperate but ultimately the government must unify and work with the IOCs to develop competency frameworks and standardisation. IOCs, such as Dana Gas and Lukoil International, have demonstrated a need and desire to improve the level of education provision in-country. Currently money is being invested in individual training facilities that prepare candidates for roles within the local operations of the company. When IOCs form a coalition to tackle workforce development and begin the process of establishing a national training centre the government will be forced to take action. The private entities with vested interests in Iraq's oil sector can, and should, take the initiative and compel the government to address the development and progress of education and training for the petroleum industry.

- The government must re-evaluate its approach to education and to the national qualifications framework. For instance, qualifications gained via e-Learning are currently not recognised. This can prevent IOCs and training providers from introducing innovative new measures to facilitate the development of the local Iraqi workforce. While the government of Iraq and the KRG both recognise that the country's TVET system is integral to economic success, very little is being done in terms of bringing together the correct resources and facilitating interdepartmental dialogue to effect change across the region.

- The exploitation of hydrocarbons has led to several large cities directly benefitting from the revenues generated. These cities have then attracted young people away from poorer rural areas. This pattern intensifies uneven distribution of wealth and, to some degree, justifies the lack of attention and investment in infrastructure beyond the cities directly experiencing benefit from oil revenues. Furthermore, the areas that most benefit are generally the seats of power for the ruling government or political authority. Thus, when the state collapses it cripples the economy. Iraq provides a good example of this phenomenon. The collapse of the Ba'athist regime led to a power vacuum and the collapse of the state along with its economy. Future hydrocarbon laws should support Iraq's nascent democracy, moving oil revenue outside of government control. Revenues should flow away from centres of power and into ministries and agencies charged with developing education and training for workforce development. Legislation must also consider the importance of the national oil company as a means through which the state manages its economic and political interests in the oil sector.

- While giving the Iraqi state-owned oil company more influence and power when dealing with IOCs is important, INOC must not manufacture job opportunities and cripple competition. Skills development should match market needs, especially in the oil industry where a market prone to volatile ups and downs can quickly change in terms of workforce needs. Support for local industry and businesses that have a stake in the supply chain of oil and gas operations is essential. Any future local content policy should introduce measures to include native equipment manufacturers. This will incentivise locals to participate in industry and create more opportunities for private sector employment.

- The Iraqi government should consider offering more favourable terms to IOCs. Historical fear should not limit what is sound business acumen. PSAs have worked for model countries, such as Brazil, where IOCs have been incentivised to meet local content requirements due – in part – to their equity stake in the hydrocarbons being extracted.

TVET remains a concern. Suggestions for the National Strategy for Education and Higher Education in Iraq, 2011–2020, include the following:

- Iraq must have a workforce development commission and there must be legislation to support the formation of such a commission.
- Public education must be decentralised and given greater autonomy to meet the needs of the oil industry and the wider economy.
- The industry has already taken steps to educate people for the oil sector, but more efforts can and must be made to forge closer ties between oil and gas operators and the education system.
- Private education providers must be involved in education provision for the industry. Private education providers can support national efforts to ready the Iraqi workforce through the introduction of innovative education delivery methods.

A Getenergy meeting in 2014, held in Abu Dhabi, gave industry figures and experts the opportunity to discuss their current and future concerns for Iraq. The following extract from the post-meeting report serves as an insight into the key issues affecting both the industry and the education sector:

> There was considerable concern at the meeting about how future skills needs could be met for the industry in Iraq as it expands. There was also a strong feeling that the conditions for greater and more effective collaboration between companies, ministries, and education and training providers should be created: the current arrangements meant that the industry made little use of Iraq's public technical and vocational education and training capacity, which, in turn, meant that the public technical and vocational institutions were largely unaware of the skills, disciplines, and future needs of the oil and gas industry.

In considering the oil and gas industry in Iraq, the Kurdistan Region (or Northern Kurdistan as it is often referred to) merits separate consideration. Whilst still a part of Iraq, the Kurdistan Region operates autonomously from national rule and the Kurdistan Regional Government has constitutionally recognised authority over the provinces of Erbil, Dohuk and Sulaimaniya. There is, at the time of writing, very little direct engagement between the machinery of government in Baghdad and the regional government administration in Erbil meaning that the growth of the energy industry – and the related issues of skills and workforce development – is being driven by a very different political, social and economic context. The Kurdistan Region will be examined as a condensed case study in the next section (following the Getenergy View).

The Getenergy View

Iraq has the chance for a new start where its oil industry is concerned. International hopes for greater, diversified energy security are pinned to the country. Our analysis of the situation in Iraq reveals the following:

- There has been a historical misalignment between the education sector and the industry. The future will require oil and gas companies to ensure education providers understand their needs in terms of skills – e.g. more technicians and skilled workers over engineers and geoscientists. This dynamic is at the heart of the challenge and it is incumbent on all protagonists to play their part. This means that the government needs to act as a convener, the international operators need to contribute in a meaningful way and the fragmented education and training sector in Iraq needs to find a collective voice. This will allow international education and training providers to build partnerships with local institutions and will begin the process of building an education and training system that is able to meet the international standards now required by the oil and gas sector.
- Local content policy should address short-term goals (e.g. industrial diversification) and long-term goals, such as the establishment of a research and development network that sits at the tertiary level of education. Local content policy should also seek to increase local participation at the supply chain level, where more jobs can be created through oil funds flowing into the country. However, Iraq must make significant strides towards driving local capacity development so that oil companies can realistically hire local supply chain companies. At the tertiary level, levies on oil companies can be channelled into universities to support strong research networks that will attract investment into the region.
- In the current environment IOCs are not incentivised to contribute towards capacity development. IOCs are contracted suppliers to the government under Technical Service Agreements (TSAs) and they do not have a long-term stake in Iraq's oil sector. TSAs relegate IOCs to consultants who are not encouraged to give good advice because they receive the same consultation fees regardless of the results.[24] PSAs are the model favoured by emerging energy nations that have secured huge gains in workforce development. The government of Iraq needs to find better ways to incentivise IOCs to meet local content requirements and invest in workforce development initiatives.
- In our view a key failing of the approach in Iraq is that there is no platform through which IOCs and government can discuss workforce development issues. The government is, at times, insular and does not provide clear guidance on what the approach to workforce development is. This has led to competing agendas between ministries and an absence of a clearly articulated, overarching

---

24. http://cabinet.gov.krd/a/d.aspx?l=12&s=02010100&r=223&a=24710&s=010000.

Continued

The Getenergy View—cont'd

plan. Improved efforts must be made to facilitate dialogue between IOCs, government and the education sector.

- The future of Iraq could be viewed as one filled with uncertainty. However, the oil and gas sector has remained surprisingly resilient despite persistent challenges to the stability of the country that have taken their toll on the local people. There is evidence that the incumbent Iraqi government has managed to create a degree of security and stability in recent years and this should increase the faith international investors have in the region. The challenge now is to give the Iraqi people a chance to feel the benefits of their oil and gas industry. This will involve building skills amongst the people and creating an industry that is run for and run by the Iraqi people.

An Extended Note on the Kurdistan Region – Iraq

A Brief Look at the Unique Factors Concerning Workforce Development for the Kurdistan Region's Oil and Gas Sector

The authors would like to express their deep gratitude to Ian McIntosh, Consultant Advisor on Local Workforce Development, Ministry of Natural Resources, Kurdistan Regional Government, Iraq for his assistance with this note.

Introduction

Despite economic concerns, political disagreements with Baghdad and the threat posed by militants in the area, the Kurdistan Region of Iraq (KRI) has managed to strike new deals for oil exploration and production and is progressing opportunities to further develop export infrastructure and continue selling its oil on the open market. This subsection will examine the Kurdistan Region's recent growth as an oil and gas producer and will provide a look towards the future of the Region and the challenges it must overcome if it is to become a strong energy economy.

Facts Related to Kurdistan

The following points relate to the KRI, its economy, oil and gas industry, labour force and other relevant information:

- The KRI first obtained its semiautonomous status in 1970 with the signing of the Iraqi–Kurdish Autonomy Agreement after the First Kurdish–Iraqi War. Four years later this agreement would fall apart and disputes would erupt into the Second Kurdish–Iraqi War.
- Years of armed conflict concluded with Kurdish Peshmerga forces pushing Iraqi forces out of Northern Iraq. By 1991, Kurdish self-rule in Northern

An Extended Note on the Kurdistan Region – Iraq—cont'd

Iraq was accepted. The KRI's semiautonomous status was upheld in the 2005 Constitution of Iraq.

- The KRI's borders are still subject to debate – especially in oil-rich areas – but some estimates indicate that if the Region was a country it would rank within the top 10 countries globally based on oil resources.[25]
- In 2013, estimates put KRI's oil resources at 45 billion barrels.[26]
- The KRI produces around half a million barrels of oil a day – representing around 15% of overall production across Iraq – production has been slowly increasing over recent years and the government has ambitions to expand production to 2 million barrels a day by 2020.
- The KRI's economy is stable and since 2003 workers have been attracted to the region from other parts of Iraq. Studies show that over 20,000 workers have relocated to Kurdistan Region from other cities in Iraq.[27]
- In 2012, the labour force was estimated to be 4.156 million.
- In January 2014, the KRG's monthly report outlined employment figures for IOCs operating in the KRI. IOCs employed 4566 individuals in the region, of which 2872 (63%) were locals.[28] No similar data for Oilfield Service Companies are available but in 2012, their workforce was three times larger than IOCs'.
- The government hopes to see infrastructure upgrades to pipelines to increase exports, primarily to Turkey. It is hoped that the pipeline commissioned in 2014 allowing direct transport of oil to Turkey will see 2 million barrels of oil exported annually by 2019.
- Infrastructure upgrades and increased production will mean workforce development becomes a crucial factor to the Kurdistan Regional Government.

KRI's Oil Sector

The nature of the oil and gas sector in the KRI is very different to the rest of Iraq in that all oil and gas activities are coordinated by the Ministry of Natural Resources (MNR) in Erbil. This has particular relevance to the nature of exploration and production contracts. Unlike the rest of Iraq – where international companies enter into service contracts meaning that they are treated as contractors by the government and the national oil company – companies operating in the Kurdistan Region are partners in production sharing agreements with the regional government. Furthermore, there is not yet a national oil company for the Kurdistan Region. The

25. "Kurdistan's Oil Ambitions". www.businessweek.com. (November 14, 2013). Retrieved November 19, 2013.
26. Bloomberg: http://www.bloomberg.com/bw/articles/2013-11-14/2014-outlook-kurdistans-oil-ambitions.
27. Barkey, H. J., Laipson, E., "Iraqi Kurds And Iraq's Future", *Middle East Policy* 12, no. 4 (2005): 66–76 [p. 68].
28. Monthly Report, Ministry of Natural Resources, Kurdistan Regional Government, Issue 4, January 2014.

Continued

An Extended Note on the Kurdistan Region – Iraq—cont'd

MNR acts as the regulator and partner in the agreements it signs with operators. These factors have created a more industry-friendly context and are driving growth across the sector.

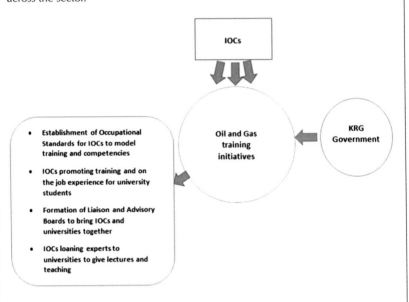

Kurdistan Region's structure of the industry.

At a production level of half a million barrels a day, the industry is comparatively small but there are around 27 operators active in the Region. Of those, 4 have direct workforce complements of over 500. A majority of the major IOCs have operations there and interest in the region has developed due to the evident opportunities that exist and the favourable exploration and operating environment. The industry is also relatively young with many (major) operators having entered the market over the last 5 years. The size and comparative youth of the sector is reflected in the current (somewhat immature) status of skills and workforce development activities. There has been, and continues to be, significant emphasis from the regional government on growing production, generating revenues and building the sector but it is only recently that the Region began to look more seriously into issues of education and training.

While Kurds did enter and work in the Iraqi oil sector during Saddam Hussein's time, this was limited and the industry within the Kurdistan Region was negligible because of economic repression. The legacy of this history is that there is relatively few senior Kurdish staff in the oil sector. This is understandable when most senior

An Extended Note on the Kurdistan Region – Iraq—cont'd

managers typically have more than 25 years experience in the industry. The KRI's oil and gas industry is a little over a decade old and consequently sourcing local senior staff is difficult. This situation has created a reliance on expatriate staff and overseas professionals.[29]

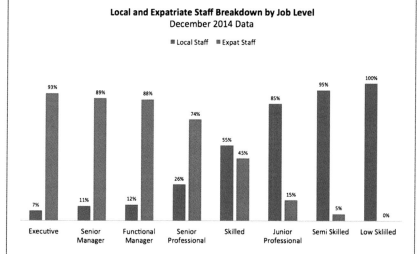

Local and Expatriate Staff Breakdown by Job Level
December 2014 Data

■ Local Staff　■ Expat Staff

Graphic provided by Ian McIntosh, Consultant Advisor on Local Workforce Development, Ministry of Natural Resources, Kurdistan Regional Government, Iraq.

Education in Kurdistan Region

In a situation mirrored in other parts of Iraq, the Kurdistan Region does have a history and heritage in higher education provision, with 11 universities in the Region currently offering relevant degree programmes in engineering and geology. These programmes produce around 300 graduates every year. However, as yet not many gain employment in the sector. This is in part due to a lack of employability skills and a disinterest amongst graduates in working in what is seen as a demanding industry. There are plans in place to double the number of graduates from these courses over coming years to meet the perceived increase in demand from the oil and gas sector. The number of petroleum engineering graduates is forecast to increase from 80 to 230 in the next 5 years. However, there are questions over the quality of the graduates currently entering the labour market (both in terms of their technical ability and suitability for the workplace) and no clear strategy is yet in place to improve programmes, outcomes and prospects for future graduates.

The challenges in relation to technical and vocational education are even more acute with no facilities currently operating effectively as training centres for oil

29. Ian McIntosh, "Who is going to manage the oil and gas industry of the future in Kurdistan".

Continued

An Extended Note on the Kurdistan Region – Iraq—cont'd

and gas technicians in the Region. This lack of provision includes non-oil- and gas-specific trades such as welders and electricians.

The ability of the Kurdistan Regional Government to invest in education and training has been dented by the political issues that exist with Baghdad, which have led to the withholding of oil revenues. This stand-off may well be exacerbated as production capability (and direct revenue generation) increases across the Region. Furthermore, if oil prices remain low, the war against insurgents in Northern Iraq continues and the value of the Kurdistan industry grows, the political dynamic with Baghdad may fracture further. Ultimately, Baghdad will have to decide whether to settle with Erbil on revenue sharing or risk conflict and a possible break-away by the Region as it seeks to capitalise on its burgeoning oil wealth.

Workforce and Skills Development for the Oil and Gas Industry

In a 2012 report produced for the Kurdistan Regional Government (KRG) employers were asked to assess the preparedness of graduates coming from secondary and postsecondary institutions. Forty percentage of respondents stated that they felt graduates from secondary and vocational institutions were 'poorly prepared' for employment.[30] Although at first glance this figure may seem negative, in many ways it is perhaps positive. It suggests well over half of the KRI's employers are not concerned with the preparedness of graduates coming from the national education system and it also suggests that there is acceptance of vocational and secondary level education and the calibre of students they produce.

The apparent confidence in the education system may be related to the progress being made, albeit slowly, to increase the quality of provision, especially university education for engineering and geoscience subjects. IOCs have begun to recognise that future exploration and production activities will be more successful if they are supported by programmes aimed at developing local talent. But roles in the oil sector are intensive and in high skill areas. Local institutions do not yet have the equipment or instructors required to develop the local workforce for these roles.

IOCs have slowly begun to respond to the situation. According to Ian McIntosh, advisor to the KRG on workforce development for the oil and gas industry, around a dozen Oilfield Service Companies and IOCs have formed Liaison and Advisory Boards with the Universities of Dohuk, Koya and Kurdistan Hawler.[31] The idea behind the advisory boards is to bring IOC staff and education professionals into dialogue, giving each the opportunity to discover the needs of the other and work together to formulate effective strategies for university capacity building.

Short-term summer placements for university students are one such solution. In 2014, almost 300 university students were given the opportunity to visit installations, work with overseas expatriate professionals, live and work along-

---

30. "An assessment of the present and future labor market in the Kurdistan Region – Iraq: implications for policies to increase private-sector employment", 2014, RAND. http://www.rand.org/content/dam/rand/pubs/research_reports/RR400/RR489/RAND_RR489.pdf.
31. Ibid.

An Extended Note on the Kurdistan Region – Iraq—cont'd

side the people responsible for the day-to-day operation of oil and gas production facilities and gain hands-on experience within the industry. Not only does this provide students with practical experience, it strengthens the relationship between students, IOCs and service companies. Other initiatives include building the curriculum capabilities of institutions, providing training equipment, software and loaning industry experts to universities as lecturers.

Within this context, the story of the KRI is one of potential. With clear plans in place to rapidly expand production and a regional government committed to creating a dynamic international oil sector, the Region is well placed to be at the heart of a new chapter in the energy story of the Middle East. And with this expansion will come the inevitable demand for skills. Whilst the regional government and the industry are currently behind the curve on this challenge, there is belief that the Region could become a new frontier for international education and training providers with real opportunities around improving teaching, facilities and curricula in the higher education sector and building internationally accredited provision in the technical and vocational sector. Much will depend on the involvement of IOCs, who have so far proven slow to respond to the challenge of capacity development due to political risk factors and a lack of revenue.

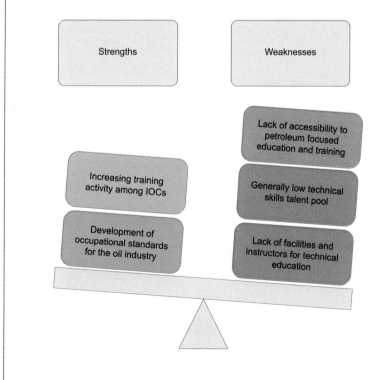

An Extended Note on the Kurdistan Region – Iraq—cont'd

The Future

The KRI is seen as a strategic area by many. It has the power to support stability in Iraq, minimise militant activity through economic growth and reduce tensions between Kurdish communities and their respective governments. For Kurdistan to prosper economically and fulfil international hopes, it must utilise its oil and gas sector for the betterment of its people. Future considerations should include the following:

- IOCs and service companies must adapt their approach to training technicians and engineers to the unique demands of Kurdistan Region. Experts suggest IOCs will have to further develop flexible on-the-job training. On-the-job training will increase participation and quicken the process of preparing professionals for the industry. The Ministry of Natural Resources is working to ensure technicians operating production facilities will demonstrate the requisite competencies. The KRG is already in the process of establishing an occupational standard for production operations that will function as a framework for IOCs seeking to educate and develop employees.
- The Kurdistan Region has a small talent pool for its oil and gas industry, but not for the same reason as many other MENA region countries. Generally, low enrolment on engineering courses or technical training is a result of entitlement cultures and a poor perception of the value of an oil and gas career. In Saudi Arabia, for example, a career in the public sector is generally favoured amongst Saudi nationals, and this has been cited as having a negative impact on interest in education for, and employment in, the oil sector. For the KRI it is the years of political and economic suppression that has caused skills gaps for the Kurds, not a low perception of the industry. This creates hope for the future.
- Traditionally the population of the Region consisted predominantly of rural workers and this has contributed to low levels of general education (particularly in the area of technical skills). Additionally, asking people from these communities to undertake work within a highly demanding industry where employees are expected to work long hours has been difficult. The IOCs and service companies will need to find flexible ways to enculturate education initiatives aimed at preparing oil and gas professionals. This is best done through creating clear incentives, such as lifelong training and well-defined career paths. Lifelong training should include soft skills as well as technical skills, giving future employees the opportunity to progress into other aspects of the business (e.g. a 'future leaders' programme).
- The Kurdistan Region will need to build on its success. Increased production will hopefully help stabilise Iraq. For the KRG, investments in education and training will ultimately lead to increased output and exports. This will help meet the government's agenda of complete economic independence. The Kurdistan Region will need the support of its neighbours (Turkey and Syria) if it is to achieve its right to full autonomy, which it has shown it is capable of through its ongoing oil trade agreements with Turkey. The KRG has already demonstrated that it has a strong democracy, a resilient economy (even if it lacks the diversification sought by other energy nations such as Brazil) and a military presence able to repel and deter militant activity. Kurdistan could

---

An Extended Note on the Kurdistan Region – Iraq—cont'd

learn some valuable lessons from Brazil in how to manage its oil sector for the benefit of the entire economy, which entails channelling contributions from IOCs' investments into education and research initiatives.

- To achieve a harmony between industrial diversification, the oil industry and education and training – and to make this a reality – the KRG may discover that establishing its own regional oil company will help. The KRG has already chosen its contractual model, PSAs, whereby the government has a stake in the operations of international operators. This model works well but it can be enhanced by state-owned oil company devoted to managing contracts to maximise local value, and able to drive forward a coherent workforce development agenda.

Glossary of Acronyms

ASET Aberdeen Skills and Enterprise Training Ltd
b/d Barrels per day
bbl/d Barrels per day
BNDES National Bank for Economic and Social Development (Banco Nacional de Desenvolvimento Econômico e Social)
CANACINTRA Cámara Nacional de la Industria de Transformación
CEFET Federal Centres for Technological Education or System S
CENAP Center for Education and Research of Oil
CENPES The Centro de Pesquisas Leopoldo Américo Miguez de Mello
CNP National Petroleum Council or O Conselho Nacional do Petróleo
CTGAS-ER Centro de Tecnologias do Gas & Energias Renovaveis (Gas Technology Centre and Renewable Energy)
DAP Discretionary allocation policy
EDTP Engineering Design Training Program
EOR Enhanced oil recovery
ETF Education trust fund
FIRJAN Federation of Industries of the State of Rio de Janeiro
IBU Iraqi Businessmen Union
IFI Iraqi Federation of Industries
IMP The Mexican Institute of Petroleum (Instituto Mexicano del Petróleo)
INOC Iraq National Oil Company
IOC International Oil Company
IPC Iraq Petroleum Company
ITF Industrial Training Fund
KMZ Ku-Maloob-Zaap
KRG Kurdistan Regional Government
LSS Local Scholarship Scheme
MoE Ministry of Education
MOHESR The Ministry of Higher Education and Scientific Research
MOL Ministry of Labour
MOP Ministry of Planning
NAFTA North American Free Trade Agreement
NDP The National Development Policy
NIS Nigerian Immigrations Service
NNPC Nigerian National Petroleum Corporation
NOC National Oil Company
NOGICD Nigerian Oil and Gas Industry Content Development Act
NQF National Qualifications Framework

OECD Organisation for Economic Co-operation and Development
PROMINP Program for the Mobilization of the National Industry of Oil and Natural Gas
PSA Production Sharing Agreement
PSC Production Sharing Contract
PTDF Petroleum Technology Development Fund
PTI Petroleum Training Institute
SABER The Systems Approach for Better Education Results
SEBRAE Brazilian Services for Assistance to Micro and Small Companies (Serviço Brasileiro de Apoio às Micro e Pequenas Empresas)
SENAC National Service for Commercial Training (Serviço Nacional de Aprendizagem Comercial)
SENAI Serviço Nacional de Aprendizagem Industrial or National Service for Industrial Training
SENAR National Service for Rural Training (Serviço Nacional de Aprendizagem Rural)
SENAT National Service for Transport Training (Serviço Nacional de Aprendizagem do Transporte)
SEP Mexican Secretariat of Public Education (Secretaría de Educación Pública)
SESC National Service for Business Training (Serviço Social do Comércio)
SESCOOP National Service for cooperative learning (Serviço Nacional de Aprendizagem do Cooperativismo)
SESI Social Service for the Industry (Serviço Social da Indústria)
SEST Social Service for the Transport sector (Serviço Social do Transporte)
SIWES Students Industrial Work Experience Scheme
Tcf Trillion cubic feet (measurement of dry natural gas)
TSA Technical Services Agreement
TSC Technical Services Contract
TVET Technical and vocational education and training
UNACAR Universidad Autonoma del Carmen
UNAM The Universidad Nacional Autónoma de México
URAP University Ranking by Academic Performance
UTs Technical universities

Index